T0135664

Diss. ETH No. 15622

Novel Perspectives of Jet-Stream Climatologies and Events of Heavy Precipitation on the Alpine Southside

A dissertation submitted to the
SWISS FEDERAL INSTITUTE OF TECHNOLOGY (ETH) ZÜRICH

For the degree of
DOCTOR OF SCIENCES

Presented by

PATRICK KOCH
Dipl. Phys. ETH
born 2 July, 1974
citizen of Villmergen AG, Switzerland

Accepted on the recommendation of
Prof. Dr. Huw C. Davies, ETH Zürich, examiner
Prof. Dr. John R. Gyakum, McGill University, co-examiner
Prof. Dr. Heini Wernli, University of Mainz, co-examiner

June 2004

Bibliografische Information Der Deutschen Bibliothek

Die Deutsche Bibliothek verzeichnet diese Publikation in der Deutschen Nationalbibliografie; detaillierte bibliografische Daten sind im Internet über http://dnb.ddb.de abrufbar.

ISBN 3-8325-0678-0

Logos Verlag Berlin
Comeniushof, Gubener Str. 47,
10243 Berlin
Tel.: +49 030 42 85 10 90
Fax: +49 030 42 85 10 92
INTERNET: http://www.logos-verlag.de

Celui qui ne comprend pas,
et qui le dit,
est celui qui fait le plus évidemment preuve
d'intelligence,
car il a compris qu'il n'a pas compris,
et c'est ce qui est le plus difficile à comprendre (...)

Albert Jacquard

Contents

Abstract

In the last two decades, several studies have emphasized the relevance of the tropopause region for the dynamics of transient synoptic weather systems. In the two independent parts of this thesis, novel perspectives are explored of two salient topics of dynamical meteorology that are related to characteristic features at tropopause level.

In the first part, the focus is on jet streams. Increased knowledge and understanding are sought related to their structure, geographical distribution, seasonal variation and dynamics. There are limitations to existing jet-stream climatologies based upon mean zonal wind distributions, since they capture only a single maximum, the so-called subtropical jet, in contrast to daily upper-air charts that reveal more intricate multiple-jet structures. In this study, a novel type of event-based jet-stream climatology is presented for both hemispheres using ECMWF reanalysis data for the years 1979-1993 (ERA15). It involves the detection of transient jet streams based upon an objective criterion. The resulting space-time distribution of jets provides a rich data set for climatological studies.
A first application is to categorize two types of jet streams based upon their baroclinic characteristics, rather than the classical subtropical/polar-front jet classification. A second application is the determination of the winter climatological positions and frequencies of single and double jets and the study of their respective dynamical characteristics with a composite approach. It is shown that single jets over the Atlantic in the Northern Hemisphere and over the western Pacific in the Southern Hemisphere have different characteristics. Moreover, the occurrence of the single- or double-jet configuration over the Atlantic significantly relates to the frequency of cyclones affecting western Europe.

In the second part, a study is undertaken of heavy precipitation episodes on the Alpine southside associated with meridionally elongated intrusions of stratospheric air (upper-level precursors) over western Europe. Particular attention is directed to the September 1993 and October 2000 episodes that caused extensive damage in the Swiss cantons of Valais and Ticino and in Northern Italy. The focus is on the relative role of the Mediterranean and the Atlantic moisture suppliers of these extreme rain events. In a sequence of numerical simulation experiments, evaporation from each water basin is removed. First, the relative impacts on precipitation in the Alpine region are investigated. It is shown that a high variability exists from one flood event to the other. In the September 1993 episode, the Mediterranean is the main supplier of moisture for the precipitation. In the October 2000 event, both water basins play a role during the first half of the event when the flow in the Alpine region is mainly southwesterly. During the second half of the episode the flow turns southerly to southeasterly and the Mediterranean has the major impact on the rainfall. Moreover, backward trajectories indicate the possible involvement of moisture of tropical origin during the second half of the event. Second, an investigation is undertaken of the impacts of the Atlantic and the Mediterranean upon the development of the upper-level precursor. It is shown that the Atlantic has the main impact and that the influence of the Mediterranean is marginal. There are nevertheless different evolutions of the upper-level precursor for the two different heavy rain episodes, when the Atlantic is removed. In the September 1993 event the upper-level precursor regresses northward. In the October 2000 event, the precursor is still present but is shifted farther eastward and has a lesser meridional extension. The differences observed in the simulated genesis of the upper-level precursor have direct implications in the determination of the regions that eventually run the risk of being affected by heavy precipitation episodes.

Résumé

Ces dernières deux décennies, plusieurs études ont souligné l'importance de la région de la tropopause pour la dynamique des systèmes météorologiques et synoptiques transitoires. Dans les deux parties indépendantes de cette thèse, de nouvelles perspectives relatives à deux sujets saillants de la météorologie dynamique et liés à des traits caractéristiques présents au niveau de la tropopause sont explorées.

La première partie se concentre sur les courants-jets. Une meilleure connaissance et compréhension de leur structure, distribution géographique, variations saisonnières et de leur dynamique sont ici recherchées. Les climatologies actuelles de courants-jets basées sur des distributions moyennes de vent zonal montrent de sérieuses limites. En effet, elles ne capturent qu'un seul maximum (assimilé au jet subtropical) au contraire des cartes météorologiques quotidiennes de la haute troposphère qui révèlent des structures plus complexes de jets multiples. Dans cette étude, une climatologie des courants-jets d'un nouveau type basée sur leurs événements et sur les données ré-analysées du CEPMMT couvrant la période de 1979 à 1993 (données ERA15) est présentée pour les deux hémisphères. Elle implique la détection de courants-jets transitoires basée sur un critère objectif. La distribution spatio-temporelle des jets ainsi obtenue fournit un riche ensemble de données pour des études climatologiques.
Une première application est la catégorisation de deux types de courants-jets basée sur leurs caractéristiques barocliniques plutôt que sur la classification classique subtropicale/front polaire. Une seconde application est la détermination des positions et fréquences climatologiques hivernales des jets uniques et doubles et l'étude de leurs caractéristiques dynamiques mutuelles en adoptant une approche composite. Il est montré que les jets uniques au-dessus de l'Atlantique dans l'hémisphère nord et au-dessus de l'ouest du Pacifique dans l'hémisphère sud ont des caractéristiques différentes. De plus, la fréquence des configurations de jets uniques ou doubles sur l'Atlantique est liée de manière significative à la fréquence des dépressions affectant directement l'ouest de l'Europe.

Dans la deuxième partie, une étude est menée concernant les épisodes de fortes précipitations au sud des Alpes associées à des intrusions méridionalement étendues dans la troposphère d'air stratosphérique au-dessus de l'ouest de l'Europe (les précurseurs de haute altitude). Une attention particulière est portée aux épisodes de septembre 1993 et d'octobre 2000 qui causèrent d'importants dégats dans les cantons suisses du Valais et du Tessin, ainsi qu'au nord de l'Italie. L'accent est mis sur le rôle relatif joué par les sources d'humidité que constituent la Méditerranée et l'Atlantique dans ces phénomènes extrêmes de pluies diluviennes. Dans une séquence de simulations numériques, l'évaporation de chaque bassin hydrologique est supprimée. Les impacts relatifs sur les précipitations dans la région alpine sont examinés dans un premier temps. Il est montré qu'une grande variabilité des résultats existe entre les deux événements. Ainsi, la Méditerranée est le principal fournisseur d'humidité pour les précipitations de l'épisode de septembre 1993. En revanche, les deux bassins hydrologiques jouent un rôle durant la première moitié de l'épisode d'octobre 2000 alors que le flux dans la région des Alpes est principalement du sud-ouest. Durant la deuxième moitié de l'épisode, le flux tourne au sud puis au sud-est et la Méditerranée a alors le principal impact sur les quantités de précipitations. De plus, des rétro-trajectoires indiquent la possible implication d'humidité d'origine tropicale durant la deuxième moitié de l'épisode. Dans un second temps, une étude est menée sur les impacts de l'Atlantique et de la Méditerranée sur le développement du précurseur de haute altitude. Il est montré que l'Atlantique a le principal impact et que l'influence de la Méditerranée est marginale. Différentes évolutions du precurseur caractérisent cependant chaque épisode pluvieux lorsque l'Atlantique est supprimé. Le précurseur régresse ainsi vers le nord lors de l'épisode de septembre 1993 alors que dans celui d'octobre 2000, le précurseur est toujours présent, bien que déplacé vers l'est et pourvu d'une étendue méridionale moindre. Les différences observées dans les genèses simulées du précurseur de haute altitude ont des implications directes sur la détermination des régions qui courent le risque d'être affectées par des épisodes de pluies diluviennes.

Chapter 1

Preamble

Nowadays, weather conditions are impacting more and more on our society. Extreme events for instance constitute a major concern as their economical (e.g. SwissRe 2003) and sociological aftermaths can have significant consequences on populations. A new concern has arised these two last decades with the recognition by the scientific community of a change in the climate associated with global warming that in turn may increase the frequency of extreme events (IPCC 2001). The improvement of the available models and concepts is therefore a continuous matter of concern for atmospheric sciences. From this perspective, the aim of meteorology is to better describe the dynamics of the atmosphere, with the ultimate aim of producing reliable forecasts for periods of time as long as possible. To this end, research has a central relevance (Fig. 1.1) and can actually be seen as an interlaced multiconnected network of four principal centres of interest:

- **Observations**: This term refers to the way we perceive the effects of weather through measurements such as soundings and space observations with satellites.

- **Data**: Measurements made in both hemispheres are circulated in an international network of weather observations constituting a large data set available for further use.

- **Modelling**: The basic equations ruling the dynamics of the atmosphere are solved numerically; the rapid development of computers has permitted the use of increasingly sophisticated numerical models that are able to represent a board range of parameters.

- **Theories**: This comprises all physical approaches and mathematical equations that quantify the atmospheric dynamics.

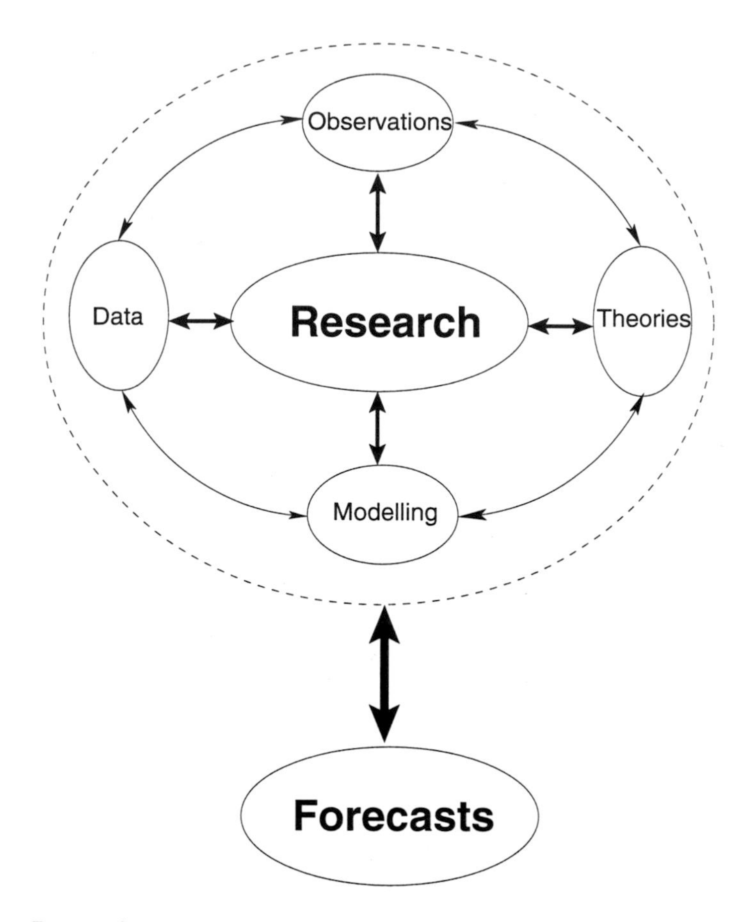

Figure 1.1: Research as a multiple inter-relationship network and its relationship to forecasts. See text for more details.

Theories permit the development of more complex models that can better simulate the behavior of the atmosphere. In turn, the results gained by these simulations trigger new questions that will result in the working out of new theories. Similar examples exist to link the other poles with one another. This whole research network is directly linked to the forecasts as it provides the forecasters with the necessary tools. Note that the experience and intuition of weather forecasters are also of primary importance. However, prediction errors still occur. The relationship between research and forecasts is then two-way as these forecast errors trigger further research in order to try and better understand the reasons for the misforecast.

This thesis is in keeping with this perspective of research as an ensemble of interacting poles. By means of state-of-the-art global data sets based upon observations, existing theories, and the latest numerical weather prediction models new perspectives are developed for two relevant aspects of dynamic meteorology.

Overview of this Study

The essence of this study is contained in Figure 1.2 which displays pressure and wind isopleths on the dynamic tropopause (defined as the 2 PV-unit surface, e.g. Chapter 2). Several synoptic-scale structures can be seen as regions with a locally low (high) tropopause characterized by a high (low) pressure whose transience directly influences day-to-day weather conditions all over the world. Inspection of Figure 1.2 reveals two interesting features. First, some regions of the tropopause are characterized by a large pressure gradient. These regions are associated with strong winds (the jet streams, letter A in Fig. 1.2). Second, a meridionally elongated region, characterized by a tropopause lower than its surrounding, is visible over western Europe (letter B in Fig. 1.2). The dynamical implications and events associated with these two particular regions constitute the features tackled in this study.

This thesis comprises two independent parts. In the first part, jet streams are studied. These tubes of strong winds, located at a height between 9 and 12 kilometers have typically a width of a couple of hundred kilometers, and can stretch over several thousand kilometers. The rapid development of the aviation industry in the last 50 years has triggered interest in a better understanding of the jet stream structure and forecast of their position. An example of the relevance of this latter point was given more recently by the meteorological support of the balloon flight around the world in 1999 (Eckert and Trullemans 1999). Note that in Europe, the public are not made sensitive to these salient features through weather forecasts broadcast on TV[1].

In the second part, studies are described relating to heavy precipitation episodes on the southern Alpine side. These extreme events are associated with meridionally elongated stratospheric intrusions over western Europe, similar to the one displayed in Figure 1.2.

Chapter 2 introduces some theoretical background as well as the different data sets and tools used in this study. Chapters 3 to 6 constitute Part I, and Chapters 7 to 10 constitute Part II, with Chapters 3 and 7 standing as introductive chapters to each Part. Finally, Chapter 11 concludes this study with a retrospective look at the results obtained in this thesis and the limitations of the methods used, as well as noting suggestions of further perspectives for the continuation of this work.

[1]Contrary to European ones, foreacasts in Northern America often supply the public with information about the jet stream. See for instance Environment Canada: http://weatheroffice.ec.gc.ca/jet_stream/index_e.html

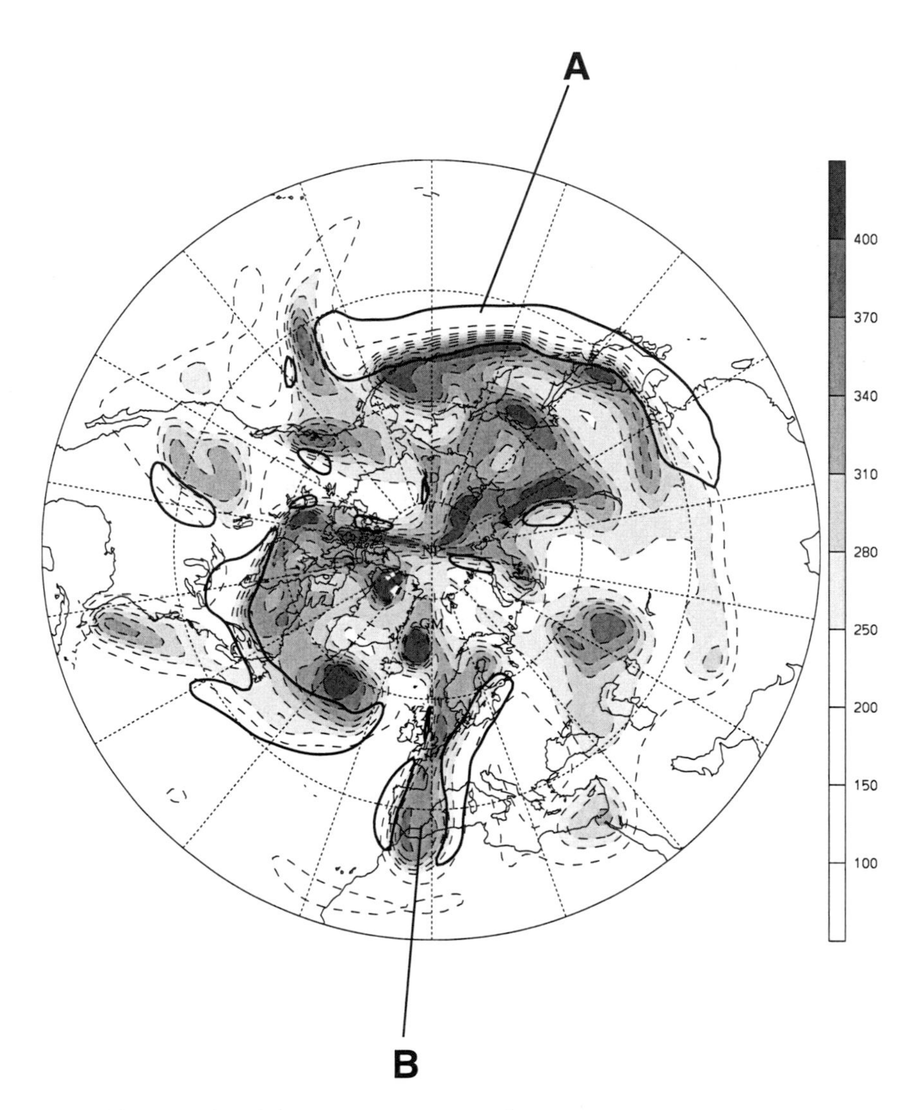

Figure 1.2: Pressure level (grey shaded, in hPa) outlined by dashed lines between 200 and 400 hPa and 40 ms^{-1} isotach (solid line) on the dynamic tropopause, defined as the 2-pvu surface at 12 UTC 13 October 2000. The letter A and B designate features discussed in the text.

Chapter 2

Physical Background, Data and Tools

This study combines the use of theoretical concepts with state-of-the-art data sets and analysis tools. In this Chapter, a brief review is first made of some basic physical aspects, followed by the description of the data sources. Finally the characteristics of the numerical model used in the simulations discussed in Chapters 9 and 10 and the method used for the trajectory calculations are summarized.

2.1 Theoretical Aspects

2.1.1 Basic Notions

A set of three equations (the so-called *primitive equations*) is used in the attempt to describe the complexity of the dynamics of the atmosphere:

$$\frac{D\vec{u}}{Dt} + 2\vec{\Omega}\wedge\vec{u} = -\frac{1}{\rho}\vec{\nabla}p - \vec{\nabla}\Phi + \vec{F} \tag{2.1}$$

$$\frac{\partial\rho}{\partial t} + \vec{\nabla}\cdot(\rho\vec{u}) = 0 \tag{2.2}$$

$$\frac{D\theta}{Dt} = \dot{\theta} \tag{2.3}$$

where $\frac{D}{Dt}$ is the time derivative following an air parcel, $\vec{u}$ is the three-dimensional wind speed, $\vec{\Omega}$ the angular speed of the Earth's rotation, ρ the mass density, p the pressure, Φ the geopotential gz, θ the potential temperature, $\vec{F}$ the non-conservative friction term and $\dot{\theta}$ the diabatic source. This set of equations comprises the momentum (2.1), continuity (2.2) and thermodynamic (2.3) equations. Note that most of the systems of equations ruling the motions of the atmosphere (i.e. geostrophic or quasi-geostrophic systems) can be derived from the set of primitive equations through suitable assumptions and scale analysis (e.g. Davies 1999).

2.1.2 Thermal Wind

The thermal wind relation is obtained by combining the geostrophic wind relation with the hydrostatic equation (Bluestein 1992, Section 4.1.6):

$$-\frac{\partial \vec{v}_g}{\partial p} = \left(\frac{R}{fp}\right)(\vec{k} \times \vec{\nabla}_p T), \tag{2.4}$$

where R is the gas constant, f the Coriolis parameter $2\Omega sin\phi$ (where ϕ is latitude), and T and p the temperature and pressure, respectively. This relation describes the link between the vertical shear of the horizontal geostrophic wind and the horizontal temperature gradient. It yields, for instance, for the zonal component of the geostrophic wind:

$$\frac{\partial u_g}{\partial p} = \frac{R}{fp}\left(\frac{\partial T}{\partial y}\right)_p, \tag{2.5}$$

i.e. a poleward decrease of temperature is linked with an increase with height of the westerly geostrophic wind component.

2.1.3 Potential Vorticity

The Ertel potential vorticity (PV hereafter) is defined as (Ertel 1942):

$$PV = \frac{1}{\rho}\vec{\eta} \cdot \vec{\nabla}\theta, \tag{2.6}$$

where ρ, $\vec{\eta}$ and θ are the density, the absolute vorticity and the potential temperature, respectively. Using the hydrostatic approximation, and casting in pressure coordinates on a spherical earth, 2.6 after scale analysis becomes:

$$PV \approx -g(\zeta + f)\frac{\partial \theta}{\partial p}, \tag{2.7}$$

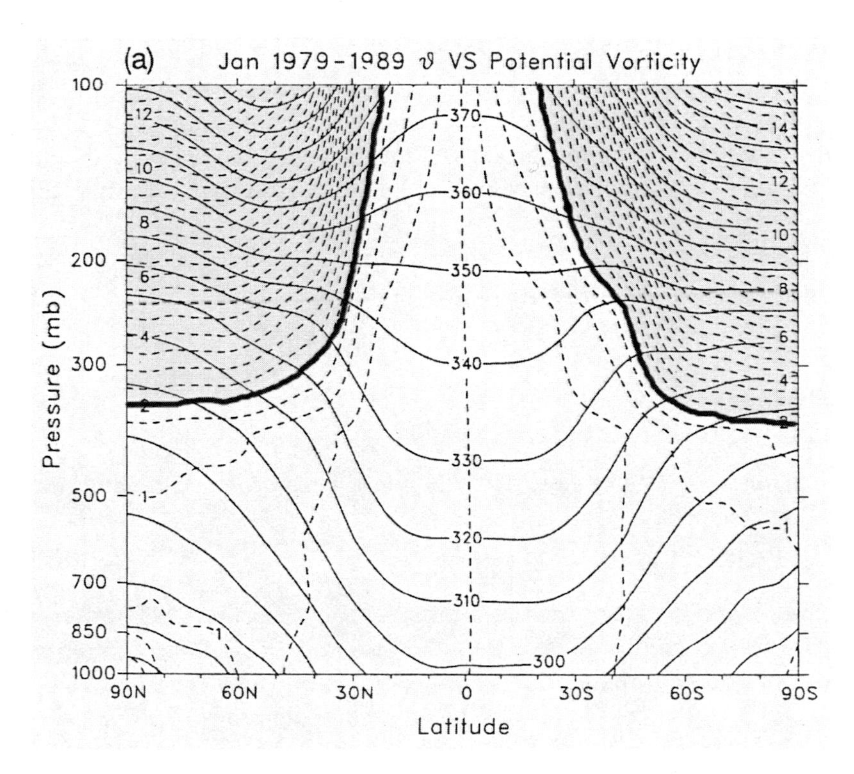

Figure 2.1: Mean January (averaged in the 1979-1989 period) vertical cross section of the zonal mean of Ertel's potential vorticity in pvu (dashed lines) and potential temperature in K (solid lines). The bold solid lines depict the dynamic tropopause (defined as 2 pvu in the Northern Hemisphere and -2 pvu in the Southern Hemisphere) and the dark grey shaded region depicts the stratosphere. Adapted from Bluestein (1993).

where ζ is the vertical component of the relative vorticity and f the Coriolis parameter. To avoid the cumbersome units of PV, PV units (pvu hereafter)[1] are used. The relations 2.6 and 2.7 indicate that PV combines a kinematic field (absolute vorticity) with a thermodynamic one (static stability). Due to the high stability of stratospheric air, climatolgical distributions of PV (Fig. 2.1) exhibit high values in the stratosphere (values larger than 2 pvu) and lower values in the troposphere (typically smaller than 1 pvu). Moreover, the β-effect also tends to distribute higher PV air in the polar regions and lower PV air in the subtropics. The dynamic tropopause (DT) separating these two regions is collocated with a zone of enhanced PV gradient on isentropic surfaces. The 2-pvu surface is quite often used to define the DT (Hoskins et al. 1985).

In the last 20 years, a large interest in the use of PV in the field of atmospheric

[1] 1 pvu is defined as $10^{-6}\mathrm{m}^2\mathrm{s}^{-1}\mathrm{K}\ \mathrm{kg}^{-1}$.

dynamics has been triggered by Hoskins et al. (1985) and the availability of large meteorlogical data sets. Moreover, the PV-θ framework (Hoskins 1991) has proved to be a powerful diagnostic tool, as the study of synoptic and global systems has been linked to the evolution and interaction of identified PV 'anomalies' (individual regions of anomalously low or high PV air compared to their surroundings)[2].
The attractiveness of the PV perspective lies in three aspects developed by Hoskins et al. (1985) and these are briefly reviewed below.

Conservation

The potential vorticity of an air parcel is conserved, provided its motion is adiabatic and frictionless:

$$\frac{D}{Dt}PV = 0. \tag{2.8}$$

In this case, PV can be regarded as a passive tracer that can facilitate the study of complex flow configurations. The adiabatic and frictionless assumption may prove to be a good approximation at upper-levels but may reveal to be harder to realize at lower-levels, where PV can be created or destroyed by diabatic sources and frictional forces:

$$\frac{D}{Dt}PV = -g\eta\nabla\dot{\theta} - g\vec{\nabla}\theta \cdot (\vec{\nabla} \wedge \vec{F}). \tag{2.9}$$

The first term on the right-hand-side of Eq. 2.9 states that a positive (negative) material PV tendency will be induced below (above) a mid-tropospheric diabatic heating release region, as suggested by Haynes and McIntyre (1987). Furthermore, the symmetry of the distribution of the diabatically induced PV anomalies depends also on the time-scales of the condensation processes and advection of the air parcels (e.g. Fig. 4 in Wernli and Davies 1997). The diabatic production of positive PV anomalies plays a major role in the formation of PV towers (Rossa et al. 2000) that are in turn relevant for the mature stage of extratropical cyclones.

Invertibility Principle

The invertibility principle (as suggested by Kleinschmidt 1950) states that the flow structure can be deduced from the PV distribution in the interior domain with appropriate boundary conditions and the specification of a reference state

[2]The use of PV by weather forecasters may also be expected to increase in the coming years.

and of a balance condition:

$$\begin{aligned} \triangle\Psi &= PV \\ \frac{\partial\Psi}{\partial z} &= \theta \\ (u, v, \theta) &= \vec{\nabla}\Psi, \end{aligned}$$

where Ψ is the streamfunction. Mathematically, this problem consists in inverting a Laplacian operator and is analogous to electrostatics (Hoskins et al. 1985; Thorpe and Bishop 1995).

The Partition Principle

The whole PV distribution can be 'decomposed' into individual PV anomalies. The individual contribution of each anomaly to the flow field can then be inferred by inverting each separately (the so-called piecewise PV inversion, e.g. Davis and Emanuel 1991 and Davis 1992), provided that a suitable linearized balance condition is used. Many studies have confirmed the utility of the piecewise PV inversion for atmospheric flow phenomena such as cyclogenesis (Davis and Emanuel 1991, Hakim et al. 1995, Stoelinga 1996), hurricanes steering flows (determined by removing their intrinsic circulation, e.g. Henderson et al. 1999, Koch 1999) or the sensitivity of Alpine rainstorms to mesoscale upper-level structure (Fehlmann et al. 2000).

2.2 Data Sets

All the data sets of prognostic meteorological fields (surface pressure, temperature, wind components and specific humidity) used in this thesis stem from the European Center for Medium-range Weather Forecasts (ECMWF). The data used in Part I (Chapters 4 to 6) are from the re-analysis (ERA) project and are available every 6 hours (00, 06, 12 and 18 UTC) for the 15-year-period 1979-1993 from the ECMWF T106L31 assimilation cycle (Gibson et al. 1997)[3]. The data used for the initialization of the numerical simulations of Part II (Chapters 9 and 10) are analysis data from the T106L31 (for the September 1993 data) and T319L60 (for the October 2000 data) assimilation cycles (see Simmons 1991 and Hortal 1991 for further information regarding the parametrization schemes used for the generation of the ECMWF data). Secondary diagnostic variables such as PV and potential temperature are derived for each type of data set.

[3]See also: http://www.ecmwf.int/research/era/ERA-15/index.html.

2.3 Analysis Tools

2.3.1 Numerical Model

The numerical simulations conducted in Chapters 9 and 10 are carried out on the Climate High Resolution Model (CHRM) that is based on the High Resolution Model (HRM) of the German Weather Service (DWD). It is a numerical weather prediction model based upon the hydrostatic set of primitive equations, using a hybrid vertical coordinate system and operating on a rotated spherical Arakawa C-grid. Following a formulation by Davies (1976), the model variables are relaxed towards their initial state values at the lateral boundaries. For this study, the model is run with 30 vertical levels.
The CHRM utilizes surface pressure, temperature, horizontal wind components, water vapor and cloud water as prognostic variables. The parameterised physical processes include a surface-layer formulation, Kessler-type grid-scale microphysics (Kessler 1969) and a mass flux convection scheme after Tiedtke (1989). Further details regarding the model set-up and the physical parameterisations, as well as previous validation studies, can be found in Majewski (1991) and Lüthi et al. (1996)[4].

2.3.2 Lagrangian Trajectories

The analysis of Lagrangian trajectories is presented in Part II. In the Lagrangian framework a physical variable is observed and characterized from a reference system moving with an elementary air volume in the atmospheric flow. The evolution of the physical properties of an air parcel can therefore be characterized along its trajectory in this framework.
In Chapters 9 and 10, the thermodynamic history of selected trajectories are calculated using the three-dimensional Lagrangian algorithm, presented in Wernli and Davies (1997) which follows a kinematic method introduced by Petterssen (1956). The required three-dimensional fields are taken from the ECMWF analysis data set. Each trajectory step ($t \mapsto t + \Delta t$) is evaluated with the following iterative scheme:

$$\vec{r}_1 = \vec{r}_0 + \Delta t \vec{u}(\vec{r}_0, t)$$
$$\vec{r}_i = \vec{r}_0 + \frac{\Delta t}{2}(\vec{u}(\vec{r}_0, t) + \vec{u}(\vec{r}_{i-1}, t + \Delta t)), \quad \text{for i} \geq 2,$$

[4]See also http://www.iac.ethz.ch/en/groups/schaer/climmod/chrm/main.html for a summary of the characteristics of the CHRM.

where $\vec{u}$ is the three dimensional wind field and the $\vec{r}_i$ are the iterative evaluations of the trajectory position at $t + \Delta t$, which are assumed to converge to the exact position $\vec{r}(t + \Delta t)$:

$$\lim_{i \to \infty} \vec{r}_i = \vec{r}(t + \Delta t).$$

Intergrid wind values are provided by linear interpolation of gridded values. Following the recommendations of Seibert (1993), a time step of 30 minutes is adopted for the iterative process.

Part I

Characteristics of Upper-Level Jet Streams

Chapter 3

Introductory Remarks

Jet streams are narrow bands of fast flowing air and are among the most salient atmospheric phenomena. In this Chapter, a short history of the discovery of jet streams is outlined and their main characteristics are reviewed. Linkages to other atmospheric phenomena are explored and the state of knowledge of multiple jets is reviewed. Finally, the aims of this study are presented.

3.1 A Short History of Jet-Stream Discovery

The attribution of the first discovery of jet streams is still a subject of debate (Phillips 1999; Krau 1999). Although observations of the motion of cirrus were conducted in England by Ley at the end of the nineteenth Century (Kington 1999), the discovery of jet streams can be attributed to Wasaburo Ooishi in Japan in the late 1924 (Lewis 2003 and references therein) through measurements of wind speed with pilot balloons. Further observations of jet streams were also conducted in Europe in the middle of the 1930s. It is most likely that the term *jet stream* itself was introduced for the first time by Seilkopf (1939) (the term *Strahlstrom* was used in German). The first in-situ experience of upper-level jet streams is attributed to the allied bombers during World War II (Barry and Chorley 2003). This empirical confirmation of the presence of jet streams at upper levels and the access to upper-level measurements then triggered a series of studies that have led to the present knowledge of the mechanisms associated with jet streams.

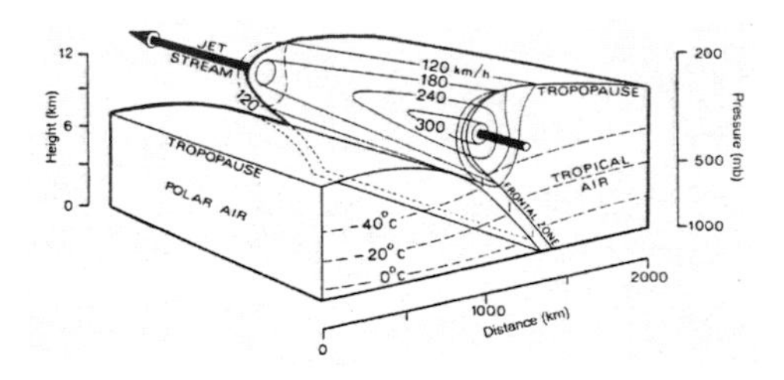

Figure 3.1: Structure of the midlatitude frontal zone and associated jet stream showing generalized distribution of temperature, pressure and wind velocity. After Riley and Spolton (1981), taken from Barry and Chorley (2003).

3.2 Key Characteristics of Jet Streams

Jet streams are narrow bands of fast flowing air that are associated with strong vertical shear (Fig. 3.1). The wind speeds and vertical shear of upper-tropospheric jets typically exceed 30 ms^{-1} and 5-10 $ms^{-1}km^{-1}$, respectively (i.e. one order of magnitude larger than synoptic-scale shear, e.g. Bluestein 1986 and references therein). The width of a jet stream is one-half to one order of magnitude less than its length.
The large vertical wind shear characteristic of jet streams is connected with regions of enhanced horizontal temperature gradient through the thermal wind relation (e.g. Eq. 2.4). Jet streams are therefore associated with fronts, which are zones of large horizontal temperature gradients. This latter fact outlines the importance of jet streams for dynamic meteorology, as meteorlogical conditions, such as temperature and winds, are expected to exhibit large variations in their vicinity. Furthermore, the identification of the jet stream and the detection of clear-air turbulence (CAT) in their vicinity (Reiter and Nania 1964) have become an important research topic linked with the development of the aviation industry (Shapiro 1976). The importance of these regions of turbulence for the mixing of tropospheric and stratospheric air and the subsequent exchange of chemical constituents has also been underlined (Shapiro et al. 1980, Shapiro 1980, Pepler et al. 1998).

3.3 Linkages to Other Synoptic Phenomena

The 'Jet Complex'

Upper-level conditions have a direct impact on the development of lower-level synoptic features. The parallel development of the theories of the relevance of upper-level divergence for cyclogenesis (Sutcliffe 1939 and Sutcliffe 1947), the advection of vorticity by the thermal wind (Sutcliffe and Forsdyke 1950), the study of confluence and diffluence regions (Namias and Clapp 1949) and the transverse circulations at the entrance and exit of the tropospheric jet (Murray and Daniels

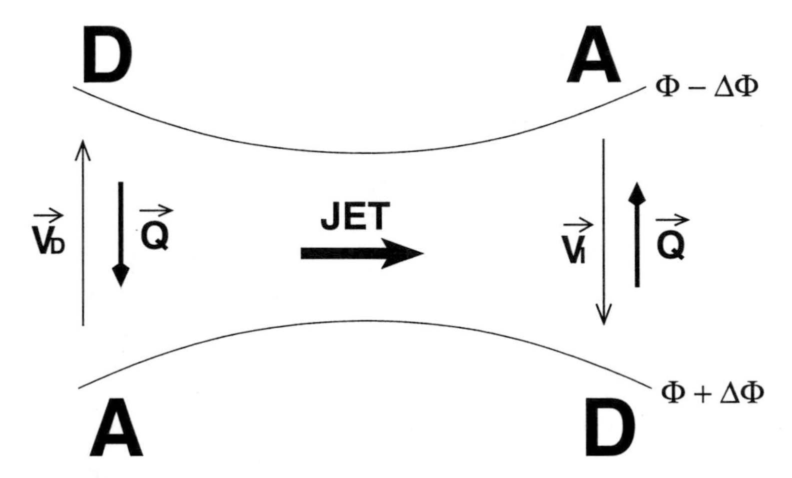

Figure 3.2: Schematic of transverse ageostrophic wind components (thin arrows, $\vec{V}_D$ and $\vec{V}_I$ representing the direct and indirect circulations, respectively), $\vec{Q}$-vectors (bold arrows) and regions of ascent (A) and descent (D) associated with the entrance and exit regions of a straight jet streak. The two arcs represent geopotential-height contours.

1953), led to the identification of regions relative to the jet where ascent or descent are favoured (e.g. Uccellini 1990 and references therein for more details). Thus ascent (descent) is favoured in the equatorward (poleward) confluent entrance and poleward (equatorward) diffluent exit regions of a straight jet. This double-dipole structure is affirmed by the $\vec{Q}$-vector perspective (Hoskins et al. 1978, Sanders and Hoskins 1990). Transverse ageostrophic wind components are associated with the douple-dipole (Blackmon et al. 1977): a direct circulation in the entrance region, directed toward the cyclonic-shear side of the jet, and an indirect circulation, directed toward the anticyclonic-shear side of the jet. Note that these ageostrophic components are necessary to offset the 'self-destructive' tendency of geostrophic flows (Davies 1999). Jet streaks are isolated isotach maxima embedded in the jet (Bluestein 1986). These jet streaks can be considered as small 'jet complexes' (e.g. the idealized jet streak case in Uccellini 1990) with the same ascent/descent structure as described in Fig. 3.2. It has been argued that these jet streaks play a role in rapid cyclogenesis (Uccellini and Johnson 1979, Uccellini 1990). Hence, the knowledge of the positions and strength of upper-tropospheric jet streams gives a direct indication about regions where ascent and cyclogenesis (or descent and anticyclogenesis) are favoured, which constitutes a most valuable asset for research and synoptic forecasters.

Jet Streams and Storm Tracks

Mobile synoptic-scale transient systems (i.e. cyclones and anticyclones) present a large variability in their genesis and lysis cycles but tend to follow climatologically preferred tracks (the so-called storm tracks). Many studies have already tackled storm tracks and essentially two main approachs have been used: (i) the track of the positions individual systems (Sinclair 1994; Sinclair 1997; Blender et al. 1997; Hoskins and Hodges 2002; Wernli and Schwierz 2004) and (ii) the analysis of Eulerian statistics of filtered variances (Blackmon et al. 1977; Trenberth 1991; Hoskins and Hodges 2002). A link exists between the climatological positions of the tropospheric jet streams and of storm tracks. As already stated in Section 3.2, jet streams are associated through the thermal wind relation with regions of enhanced horizontal temperature gradients (i.e. baroclinic zones) that are in turn favourable for cyclogenesis.
However the relation between tropospheric jet streams and storm tracks is not ubiquitous. In effect, cyclogenesis can be favoured by mountain ranges (lee cyclogenesis) and also by the temperature contrast between land and water masses without necessarily the presence of a jet stream aloft. Moreover, some jet streams appear not to be associated with a baroclinic zone that extends to the lower-levels (e.g. Section 3.4), which would set the favourable conditions for surface cyclogenesis. Nevertheless, several studies have recently focussed on the possibility to link the climatological variability of the strength, width and position of tropospheric jet streams with the variability of the storm tracks[1] (Nakamura 1992; Harnik and Chang 2004; Nakamura and Shimpo 2004).

3.4 Multiple-Jet Configurations

Subtropical and Polar-Front Jets

The study of the global terrestrial circulation in the first part of the twentieth century led to the elaboration of the three-cell meridional circulation scheme (Rossby 1939) and to the distinction between two types of tropospheric jet streams (e.g. Chapter 4 in Palmén and Newton 1969 and references therein):

- The polar-front jet (PFJ)
- The subtropical jet (STJ)

[1]In these studies, storm tracks are defined as regions of strong high-frequency fluctuations associated with synoptic-scale baroclinic waves.

Subtropical Jet	Polar-Front Jet
Wintertime phenomenon	Annual phenomenon
Jet core at nearly 200 hPa	Jet core at nearly 250 hPa
Latitude between 20° and 35°	Latitude poleward of 40°
Poleward branch of the Hadley cell	Linked to the midlatitude polar front
Shallow layer meridional temperature gradient	Deep layer meridional temperature gradient
Continuous band structure at nearly constant latitude	Disrupted meridionally meandering structure

Table 3.1: Summary of the respective characteristics of subtropical and polar-front jets as found in Bluestein 1993 (pp. 378-390).

Although they are both related to a change of height of the tropopause (and subsequently to temperature gradients) each type of jet has its own position and dynamical characteristics. Bluestein (1993) has summarized the major characteristics of both types of jet (e.g. Table 3.1). Winter subtropical jets in the Northern Hemisphere have been studied by Krishnamurti (1961), who found that they were quasi-steadily centered around 27.5°N. They are associated with a shallow layer meridional temperature gradient located between the middle and upper-troposphere. According to Bluestein (1993), the subtropical jet disappears in summer so that only the polar-front jet remains at midlatitudes. Conversely, polar-front jets are associated with polar fronts (Bjerknes and Solberg 1922) and baroclinicity extending down to lower-levels. It appears therefore that by knowing the mean position of each type of jet, one can have an idea of the mean vertical baroclinic structure of the atmosphere.
These definitions of the subtropical and polar-front jets do not lend themselves to the process of distinguishing between each type of jet because of their inherent arbitrary nature, especially when the polar-front and subtropical jets tend to be vertically superposed in the case where they migrate equatorward and poleward, respectively (Reiter and Whitney 1969).

Multiple-Jet Occurrences

Flow configurations exist where two or more jets can be found meridionally distributed on certain longitudes (i.e. multiple jets) as proposed by Shapiro et al. (1987) (Fig. 3.3). A double-jet structure is frequently found in winter in the Southern Hemisphere around 180°E and the transition between the single and double-jet configurations has been already studied (Yoden et al. 1987; Chen

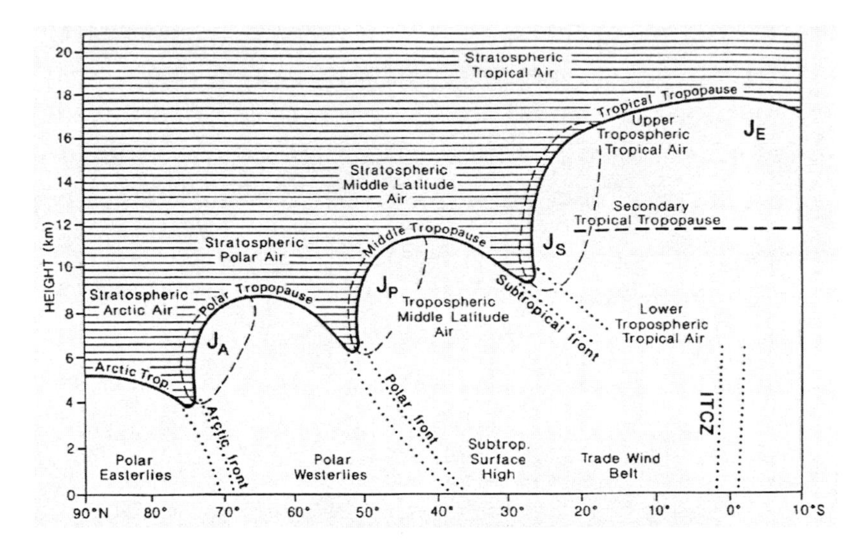

Figure 3.3: Meridional structure of the tropopause. The dynamic 2-pvu potential-vorticity tropopause is shown by the heavy solid line and the stratosphere is stippled. The primary frontal zones are bounded by heavy dotted lines and labelled accordingly. The 40 ms^{-1} isotach (thin dashed line) encircles the core of the three primary jet streams: arctic Ja, polar Jp and subtropical Js. From Shapiro et al. (1987).

et al. 1996; Bals-Elsholz et al. 2001). Double-jet configurations can also be found in the Northern Hemisphere and their position relative to each other has been shown to play a role in the occurrence of severe tornadic storms in spring in the Midwest when the left exit region of the southernmost jet comes in the vicinity of the right entrance region of the northernmost jet (Whitney 1977).

Studies have demonstrated that a relatively small meridional barotropic wind shear superimposed upon a symmetric zonal flow can lead to very different life cycles (LC) of extratropical cyclones (Hoskins and West 1979, Hoskins and Valdes 1990, Davies et al. 1991). Three different life cycles were proposed depending on the nature of the barotropic shear (neutral, cyclonic or anticyclonic, e.g. Shapiro and Keyser 1990, Wernli 1995). In his simulations, Wernli (1995) controlled the type of baroclinic shear through the adjustment of a shear parameter. More recently, Shapiro et al. (1999) proposed a 'jet perspective' of the three cyclone life cycles. They performed simulations of actual baroclinic life cycles and proposed a classification based on the position of the developing cyclone relative to the upper-level jets. Thus the nonshear (LC1), cyclonic shear (LC2) and anticyclonic shear (LC3) cyclones are located beneath the vertically aligned polar and subtropical jets, north of the subtropical jet, and south of the subtropical jet (Fig. 3.4), respectively. At upper levels, PV wave life cycles can also be reproduced by imposing a small barotropic shear to the zonal symmetric flow. The baroclinic wave life cycle associated with the LC1 (no shear) and LC2 (cyclonic shear)

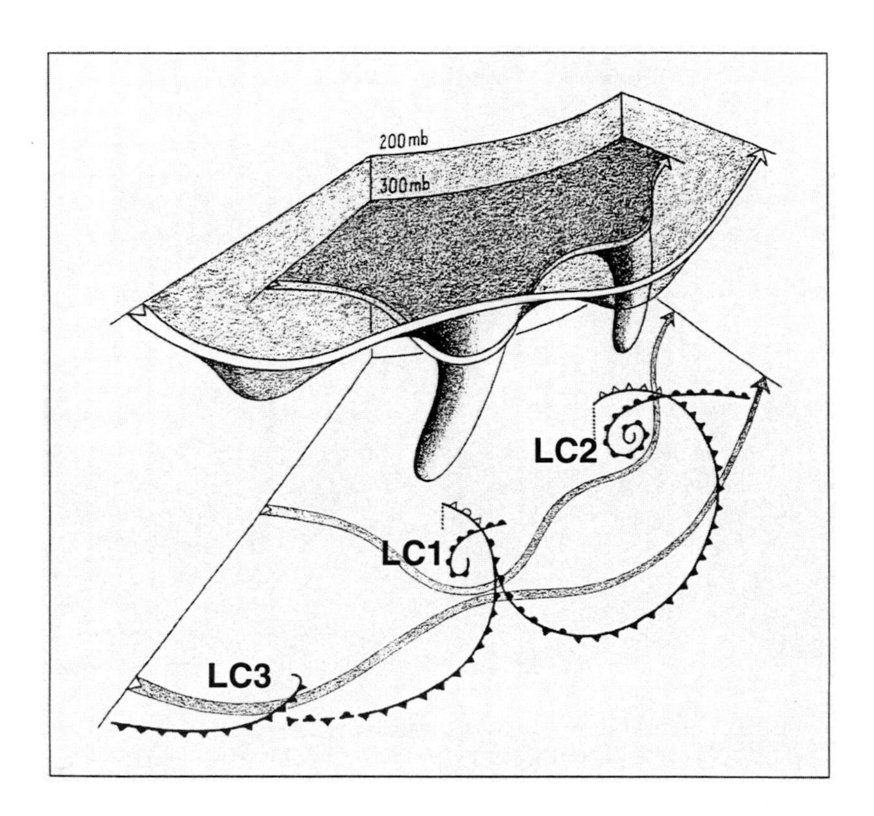

Figure 3.4: Conceptual hypothesis for the influence of upper-level jet stream and PV alignments on frontal structure within extratropical cyclones. Upper-plane (light shading): 200-mb planetary wave and subtropical jet stream (white ribbon) with associated PV anomalies suspended below. Middle plane (heavy shading): 300-mb synoptic wave and polar-front jet stream (white ribbon) with associated PV anomalies suspended below. Lower plane: Earth's surface with three characteristic cyclone frontal configurations (LC1, LC2 and LC3); frontal symbols are conventional with fronts aloft entered as open symbols. Adapted from Shapiro et al. (1999).

categories have been studied in detail by Thorncroft et al. (1993).
Multiple jets are a very complex topic in fluid dynamics and many studies have tackled the difficult task of simulating multiple-jet regimes using idealized models and the transition from one mode to another. Most of these studies are based on the works of Rhines (1975) and Williams (1979b, 1988). Panetta (1993) studied the existence, spatial persistence and meridional structure of low frequency zonal jets with a two-layer quasigeostrophic (QG) symmetric model on a beta-plane. He found jets to be flanked by a pair of storm tracks and that there was a high sensitivity of the number of jets to the size of the domain. He also pointed to limitations in the comparison with the (non-symmetric) troposphere. Lee (1997) used the same model to study transitions from the single-jet state to the double-

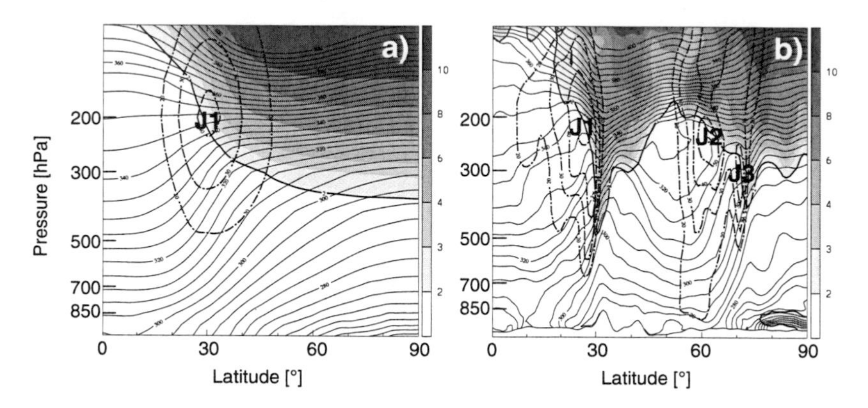

Figure 3.5: Latitude-height cross sections of PV (shaded, in pvu, the 2-pvu line being outlined by a solid bold line), zonal wind flow (dash-dotted line, contours are 20, 30, 40, 50 and 60 ms^{-1}) and potential temperature (solid lines, contours every 4 (LC) K). The vertical coordinate is logarithmic in pressure (in hPa). Panel (a) depicts the wintertime (DJF) longitudinally average distribution for the ERA15 period. Panel (b) depicts the vertical cross section at 5°W on January 27th 1989, 12 UTC. J1, J2 and J3 indicate jet cores.

jet state and the possible influence of the subtropical jet upon the polar-front jet has been studied by Lee and Kim (2003).

3.5 Aims of Part I

In this Chapter, the relevance of the tropospheric jet streams for the study of synoptic meteorology has been outlined through their linkage to other synoptic phenomena and the subsequent need for a good knowledge of their structure, geographical distribution, seasonal variation and dynamical characteristics is evident. However, many aspects of these localized regions of strong winds still remain to be studied.

Nowadays jet-stream climatologies still do not render a satisfactory picture of the phenomenon. As a matter of fact, available climatological mean zonal wind distributions for the Northern Hemisphere (Fig. 3.5 a) exhibit a single maximum, the so-called subtropical jet, associated with the corresponding space-time average climatological tropopause break. In contrast, daily upper-air charts may reveal more intricate multiple-jet structures, including extratropical jets associated with *in-situ* transient synoptic-scale weather systems (Fig. 3.5 b). There is then a need for a new type of climatology that could associate the large view of a 'classical' climatology with the transient characteristics of synoptic-scale weather systems.

The distinction between subtropical jets and polar-front jets implies a large part of arbitrariness and is of practical use on weather maps. In effect, many textbooks use definitions comparable to those given in Table 3.1 and no study, to the knowledge of the author, has been conducted to better characterize the properties and seasonal evolutions of each type of jet. There is therefore a need for a new classification of jet streams (comparable to the subtropical/polar-front distinction) based on objective dynamical criteria.
Earlier studies suggest that single or double jets occur in very specific regions. However, no study has been conducted to determine precisely these regions, nor has it been determined how often each type of jet configuration occurs and the general meridional position of the jets when a single or a double jet occur. Moreover, the occurrence of single- or double-jet configurations in a given region implies different characteristics of the dynamical properties of the tropopause. It is then interesting to characterize the typical states of the troposphere in single- and double-jet configurations and to compare the situations in the Northern and Southern Hemispheres. The aims of Part I are:

- to develop a new type of jet climatology based upon events, to capture jet events and to further investigate their seasonal characteristics
- to dynamically distinguish between different types of jet (e.g. subtropical from polar-front jets)
- to isolate single-jet and double-jet events and to build a composite of the troposphere when both spatially and temporal coherent jets of each type of configuration occur in specific regions

In Chapter 4, the methodology used for the identification of the jet events is presented and seasonal charts are analyzed. The interannual variability of the jet events is also discussed. Chapter 5 focusses on the determination of the baroclinic characteristics of the jet events and presents an objective method to distinguish between different jet types. In Chapter 6 the occurrence of single-jet and double-jet configurations is studied and composites are realized for each type of jet configuration in order to study their associated mean spatial and temporal characteristics.

Chapter 4

An Event-Based Jet-Stream Climatology

In this Chapter a novel type of jet-stream climatology is presented. It depicts the geographical distribution of jet-stream frequency. The occurrence of a jet is ascertained by evaluating the vertically-averaged horizontal wind velocity in the 100-400 hPa layer. Time and positions are recorded when the vertically averaged velocity is larger than a given threshold. Seasonal climatological frequencies are presented for both hemispheres using a 15-year data set. Changes in amplitude and structure of the frequency patterns are studied during the major interannual flow variations.

4.1 Data Set and Jet-Stream Identification

Global weather data with high temporal and spatial resolution are required to carry out the jet analysis. In this study, the six-hourly ERA-15 reanalysis data available from the ECMWF are used (see Section 2.2). For the present study, they are interpolated on a regular 1° × 1° longitude/latitude grid.

Establishing the occurrence of a jet event is realized in two steps. First the horizontal wind velocity is computed and vertically averaged in the 100-400 hPa layer at each grid point and for each six hourly time instance:

$$\alpha VEL = \frac{1}{p_2 - p_1} \int_{p_1}^{p_2} (u^2 + v^2)^{1/2} dp \tag{4.1}$$

where u and v are the zonal and meridional wind components, and p_1 and p_2 the highest (100 hPa) and lowest (400 hPa) pressure levels of the considered layer, are respectively. The use of the horizontal distribution of αVEL enables the direct

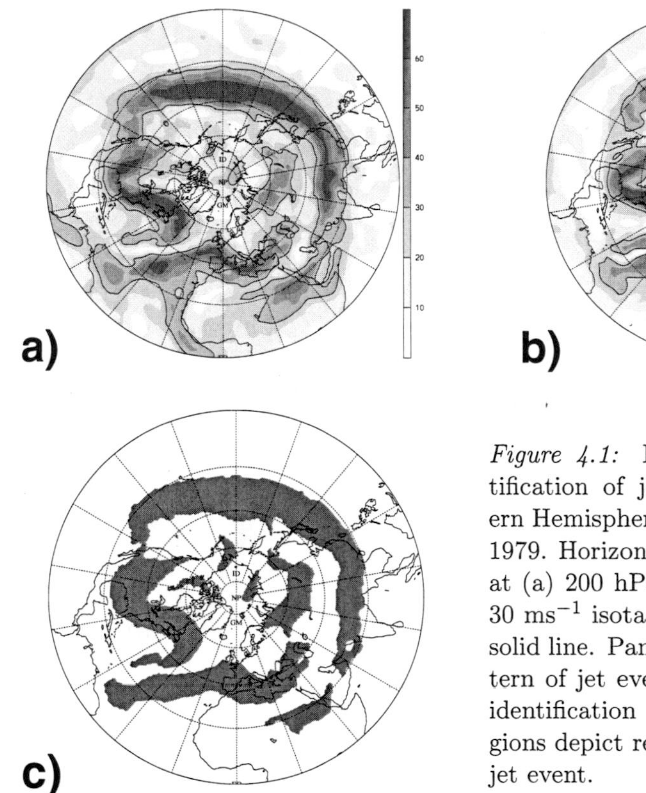

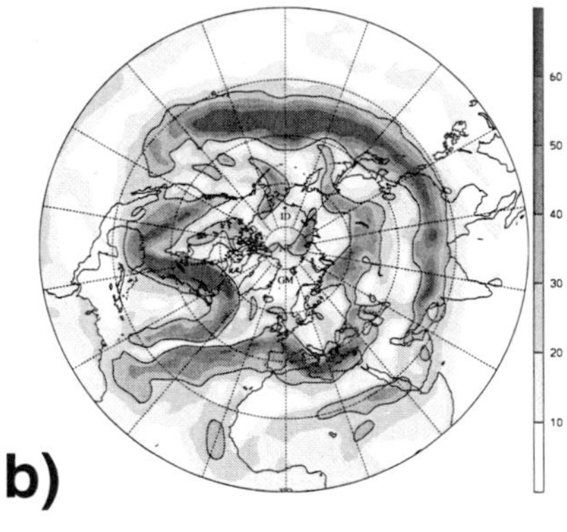

Figure 4.1: Illustration of the identification of jet events in the Northern Hemisphere for 00 UTC, 3 January 1979. Horizontal wind velocity in ms^{-1} at (a) 200 hPa and (b) 300 hPa. The 30 ms^{-1} isotach is outlined by a black solid line. Panel (c) illustrates the pattern of jet events as diagnosed by the identification method. Dark gray regions depict regions in the vicinity of a jet event.

capture and compact measure of the intensity and vertical structure of jet cores not necessarily located at the same height. A similar method was previously used by Massacand et al. (1998) for the study of PV streamers and cut-offs. Second, a search is made (for all time instances) of all locations with values of the scalar αVEL larger than a specific threshold value. Each location fulfilling this criterion is then identified as being in the vicinity of a jet-stream event and the time instance and location are recorded. At each time instance the pattern of 'events' thus defined forms disjoint meandering structures around the globe. Figure 4.1 illustrates the method by showing the horizontal wind velocity on the 200 and 300 hPa surfaces and the derived pattern of jet events for a particular time instance. This study focusses on jet streams located in the vicinity of the tropopause. The choice of the 100-400 hPa layer is therefore motivated by the fact that the tropopause is usually found in this layer outside the tropics (e.g. Fig. 2.1). Note that another possibility would have been to determine wind speed on the dynamic tropopause (usually defined as the 2-pvu isosurface). However, jet streams are located in regions of tropopause breaks and folds. Jet events in the vicinity of this latter feature would have raised questions about the physical interpretation to give to the results. The resulting climatology is sensitive to the

choice of the layer in which αVEL is calculated, the resolution of the grid on which data are interpolated and the wind speed threshold criterion used to identify jet events. An illustration of the sensitivity of the jet event patterns to the threshold criterion can be found in the Fig. A.1 of the Appendix for the event of Fig. 4.1. In our case, we are particularly interested in jet events occuring in the subtropics and extratropics. A speed threshold criterion of 30 ms^{-1} (108 km/h) tends to successfully define and separate jet streams in these regions and this wind speed will be adopted for the threshold criterion in the rest of this study.
The time-mean frequency of the jet events is calculated for each of the 15 $\times$ 12 months of the ERA15 period for the threshold criterion of 30 ms^{-1}. Individual monthly and seasonal[1] mean geographical distributions of the jet events are eventually determined.

4.2 Seasonal Climatologies

In this section the seasonal mean climatological distributions of the jet events are presented for both hemispheres. A particular attention is given to the structure and amplitudes of the frequencies of the jet events. Note that the frequencies are given in % such that a frequency of 60% implies that a jet event has been found in 60% of the total time instances of the considered month/season.

4.2.1 Northern Hemisphere

The winter (DJF, Fig. 4.2) climatology of jet events shows a spiral-like pattern with one end over the eastern Atlantic south of 20°N and the other one some 30° poleward over Great Britain. The intensity within the spiral is inhomogeneous with regions of higher or lesser frequencies. Regions that extend from eastern Asia to the mid-Pacific and from eastern North America to the western Atlantic are characterized by values exceeding 90% and 75%, respectively.
Spring (MAM, Fig. 4.2) is characterized by two regions of high jet event frequencies. One branch spirals poleward from the eastern Atlantic north of the Cape Verde Islands to the western coast of the United States and a second one extends from Hawaii to the southern tip of Greenland. Regions of enhanced frequencies are localized over southern Libya, Egypt and the Arabian Peninsula (with values

[1]Throughout this study, the four seasons are defined for the Northern Hemisphere as December-January-February (DJF) for winter, March-April-May (MAM) for spring, June-July-August (JJA) for summer and September-October-November (SON) for autumn. For the Southern Hemisphere DJF is used for summer, MAM for autumn, JJA for winter and SON for spring.

exceeding 60%), between Japan and the central Pacific (with values that exceed 75%) and over the southern United States (values over 50%).
The summer frequencies (JJA, Fig. 4.2) exhibit a meridionally thin incomplete ring-like pattern of low frequencies with one end over Turkey and the other one over Great Britain. Local regions of slightly enhanced frequencies (exceeding 30%) are distributed over the Black Sea, central Asia, the Pacific and the eastern North American/western-Atlantic sector.
Autumn (SON, Fig. 4.2) is characterized by an incomplete spiral-like pattern with a distinct branch of low frequencies over the Sahara. Two regions of enhanced frequencies are found over Japan and the western Pacific (values that exceed 75%) and between the Great Lakes and the mid-Atlantic (values over 60% over Nova Scotia).

4.2.2 Southern Hemisphere

The southern winter pattern of jet event frequencies (JJA, Fig. 4.3) is characterized by a compact spiral-like pattern extending from south of the Reunion Island to the area south of New Zealand. Frequencies are quite large inside the spiral (e.g. the 50% frequency that is nearly the spine of the spiral pattern). Two regions of enhanced frequencies are identifiable: one extending in the 20°S-40°S belt between the mid-Indian Ocean and the eastern Pacific area (with values exceeding 90% in a small region south of New Caledonia) and a second one lying poleward of 35°S between the western Atlantic and the eastern Indian Ocean (with maximum values that exceed 75% over the mid-Indian Ocean). There is a region of relatively low frequencies (not reaching 30%) present between Tasmania and the mid-Pacific.
Two main patterns are visible in spring (SON, Fig. 4.3). A first one is constituted by a zonally elongated band in the 20°S-40°S belt, extending between the eastern Indian Ocean and the mid-Pacific. A second pattern is a meridionally irregular ring (thinner over the Pacific) that encloses Antarctica. Regions of enhanced frequencies are found over western Australia and the western Pacific (over 60%) and between the mid-Atlantic and the eastern Indian Ocean, with maximum frequencies of over 75% southeast of South Africa. There is also a region of low frequencies between Tasmania and the mid-Pacific.
The main feature for the summer frequencies (DJF, Fig. 4.3) is a ring-shape pattern around Antarctica poleward of 40°S. Individual regions of enhanced frequencies are found west of Patagonia (over 50%) and between the mid–Atlantic and the eastern Indian Ocean (values exceeding 60%).
A ring semi-shaped pattern is visible poleward of 40°S enclosing Antarctica between Patagonia and the mid-Pacific in autumn (MAM, Fig. 4.3), as well as a small zonally elongated region from north of New Zealand to the western Pacific.

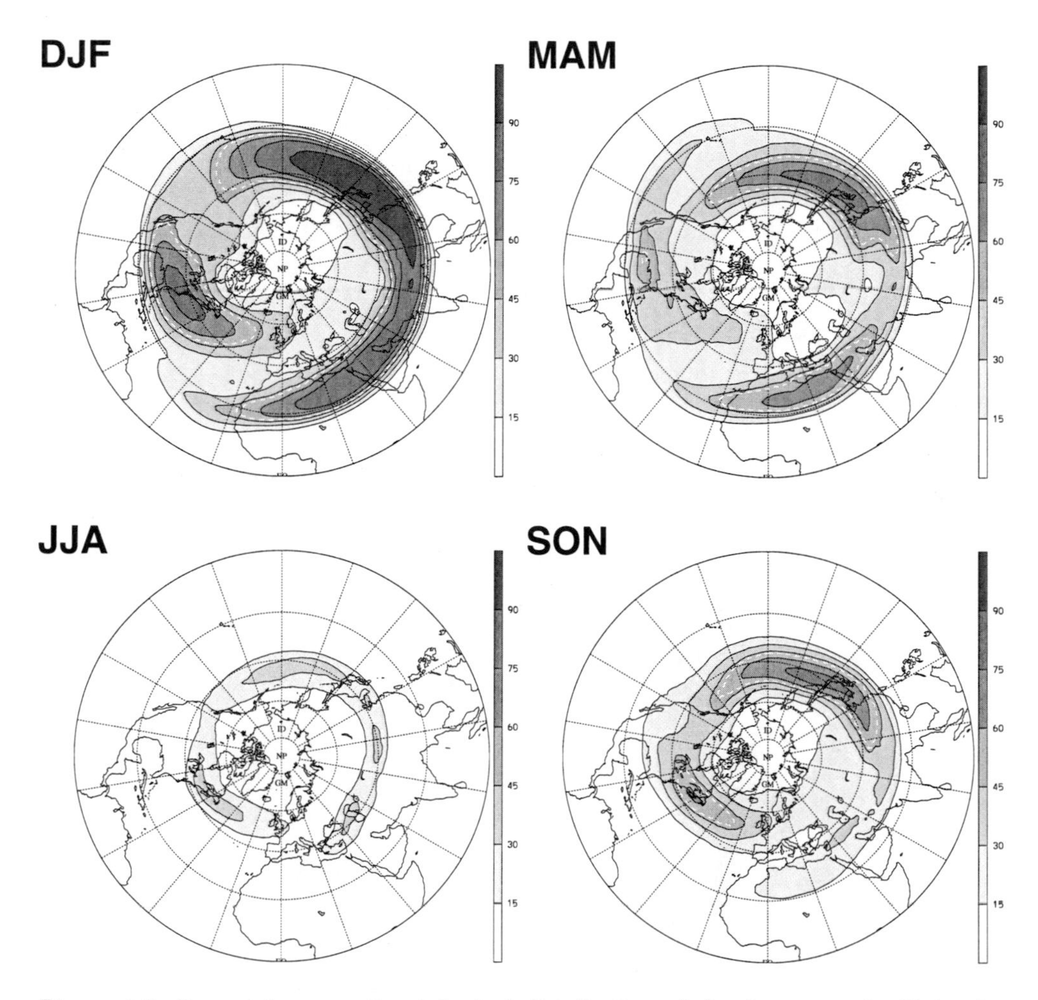

Figure 4.2: Seasonal mean climatological distribution of the jet events in % in the Northern Hemisphere for the whole ERA15 period, the dashed white line denoting 50%. Seasons are winter (DJF), spring (MAM), summer (JJA) and autumn (SON).

Regions with enhanced frequencies are found north of New Zealand and in the western Pacific (over 50%) and extending between the Atlantic and the eastern Indian Ocean (values exceeding 60%).

4.2.3 Discussion

Winter jet frequencies in the Northern Hemisphere are characterized by a spiral-like pattern of heterogeneous intensities (Fig. 4.2). This spiral structure is a recurrent feature of jet studies. For instance, Zillig (2002) found a similar pat-

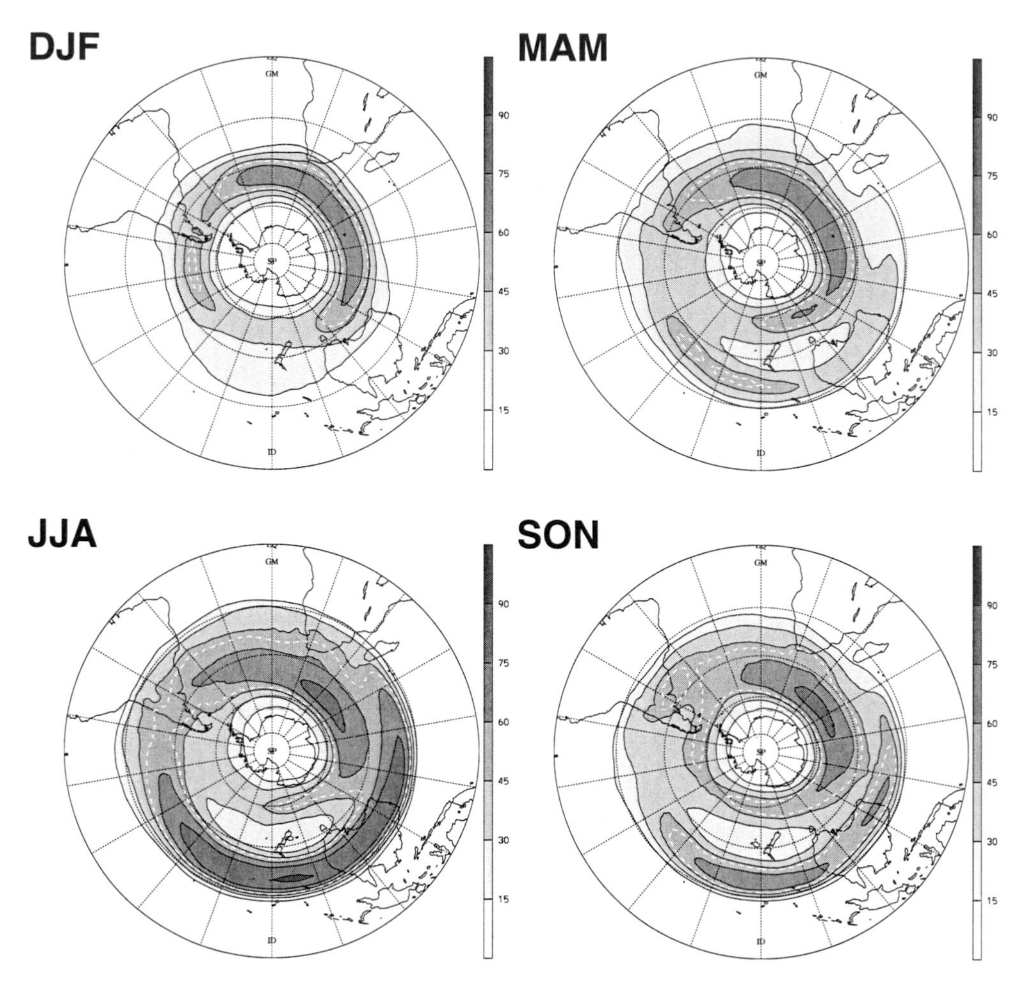

Figure 4.3: Same as Fig. 4.2 but for the Southern Hemisphere. Seasons are summer (DJF), autumn (MAM), winter (JJA) and spring (SON).

tern for the distribution of 2-pvu contours on different isentropic surfaces in his study of long straight zonal jets. The shape and amplitude of the spiral pattern has a large seasonal variability. It evolves from the winter spiral, meridionally extending to the subtropics, to a shrunk spiral with much lower frequencies and located poleward of 35°N. Conversely, winter in the Southern Hemisphere is characterized by a more compact spiral pattern and by the omnipresence of a ring-like pattern of jet frequencies that surrounds the whole hemisphere in the 40°S-60°S belt during the whole year, meaning that jet events are found throughout the year in this region. The spiral pattern is then formed by the development of a branch of jet frequencies located between the Indian Ocean and the Andes between autumn and spring.

Jet events are not equally distributed in both hemispheres and they tend to occur more often in certain particular regions. In winter in the Northern Hemisphere, two main areas are found where jet events occur most frequently. A first region is delimited by a long band extending roughly from the coasts of Mauritania to north of Hawaii in the 20°N-40°N belt, with near omnipresence of jet events from the eastern Himalaya to the mid-Pacific, as shown by the frequencies that exceed 90% in this region. A second region extends from Baja California to the mid-Atlantic with very common jet events from southeastern United States to the western Atlantic (frequencies exceeding 75%). In spring, the frequencies of jet events decrease but are still large over the Sahara and Arabian Peninsula and between the eastern Himalaya and the eastern Pacific. Jet events occur much less over eastern North America, in the Himalayan region, and from Hawaii to the western coasts of the United States. The summer is characterized by a general low occurrence of jet events, while in autumn they are a common feature over Japan and the western Pacific and between southeastern Canada and the western Atlantic.

Large wintertime frequencies of jet events are wider spread in the Southern Hemisphere than in the Northern Hemisphere. However, there is no large region of very high frequency in the Southern Hemisphere analogous to the eastern Asia/western Pacific area in the Northern Hemisphere. Two distinct regions with high jet event probabilities are found extending from the Indian Ocean to the mid-Pacific in the 20°S-40°S belt and extending from the western Atlantic to the eastern Indian Ocean in the 40°S-60°S belt. The inter-seasonal variation of the amplitude of the jet frequencies is much smaller in the Southern Hemisphere. Two particularities in the frequency of jet occurrences are visible all year round in the Southern Hemisphere. First, jet events are quite often found in the 40°S-60°S belt between the mid-Atlantic and the eastern Indian Ocean, as expressed by the quasi-steady zone of frequencies exceeding 60%. Second, New Zealand is located in a region characterized by low frequencies of jet occurrences throughout the year. This region is, both poleward and equatorward, surrounded by two areas with higher frequencies from autumn to spring.

Finally, note that monthly jet event climatologies have also been calculated for both hemispheres and can be found in the Figs. A.2 and A.3 of the Appendix. They give further details for the annual evolution of the pattern of jet events, as for the spring evolution of the jet frequencies over the eastern Pacific and the appearance of the two-band pattern of frequencies in the Northern Hemisphere, as well as the evolution from a ring-shape to a spiral-shape of the pattern of jet events in the Southern Hemisphere climatology.

4.3 Interannual Variations

Interannual variations of the global atmospheric circulation are a common feature in both hemispheres. They are more noticeable in winter when the general circulation is fully developed. The different modes of flow variability are quantified by indices based upon temperature, geopotential or sea-level pressure (Rogers and van Loon 1982, Hurrell 1995, Thompson and Wallace 1998). In this Section, interannual variations are presented of the winter jet event climatology. In particular, the variability of the jet event climatology patterns is studied for five major atmospheric flow variability patterns: the Northern Atlantic Oscillation (NAO), the Pacific North American pattern (PNA) and the Arctic Oscillation (AO) in the Northern Hemisphere and the Southern Oscillation (SO) and Antarctic Oscillation (AAO) in the Southern Hemisphere. For each type of oscillation, separate event climatologies are calculated for the months belonging to each flow mode (e.g. positive or negative indices). Note that very few months can present a neutral flow mode (e.g. a zero index). These months are not taken into account in the present study. The study uses monthly values of the different indices of the Climate Prediction Center (CPC)[2], except for the monthly NAO values that are derived from Jones and Wheeler (1997)[3].

4.3.1 Northern Hemisphere

North Atlantic Oscillation

The Northern Atlantic Oscillation (NAO hereafter) is a 'seesaw' variation of atmospheric mass between the polar and subtropical regions that is affecting the North Atlantic-European region (Wallace and Gutzler 1981). The NAO index is based on the normalized sea-level pressure difference between the Subtropical (Azores) high and the Subpolar (Iceland) low (e.g. Hurrell 1995, Barry and Chorley 2003, Chapter 7). Dynamically, the positive NAO phase corresponds to a stronger than usual subtropical high and a deeper than normal Icelandic low. This synoptic configuration is linked to a more northerly position of the jet stream and of the Atlantic storm track. The negative phase, on the other hand, is characterized by a weak subtropical high and Icelandic low and a more zonally aligned jet stream and storm track.[4]
There is no evident difference between the NAO phases apart from the region of very low jet event frequencies (less than 15%), extending over half of the Atlantic, southwest of Spain, and the slightly extension farther upstream of the

[2]e.g. http://www.cpc.ncep.noaa.gov/data/indices
[3]See also http://www.cru.uea.ac.uk/cru/data/nao.htm.
[4]e.g. also http://www.ldeo.columbia.edu/res/pi/NAO/

equatorward end of the spiral (Figs. 4.4 a and b). However, the difference in the amplitudes of the jet event frequencies inside the spiral (Fig. 4.4 c.) shows a tripole structure over the Atlantic with two positive regions located poleward of 40°N (intensity exceeding 25%), and at about 20°N (intensity not exceeding 15%), and a negative region extending farther downstream to Turkey (with maxima over 20% over the mid-Atlantic and the eastern Mediterranean). Outside the Atlantic region, other differences (smaller in intensity) are found between the Sahara and the Arabic peninsula, over the Arabian Sea and the Iran/Pakistan sector (a dipole structure with the largest intensity not exceeding 15%). Note also a small zone of lesser jet event frequencies (10% less) in the NAO- phase over the central United States.

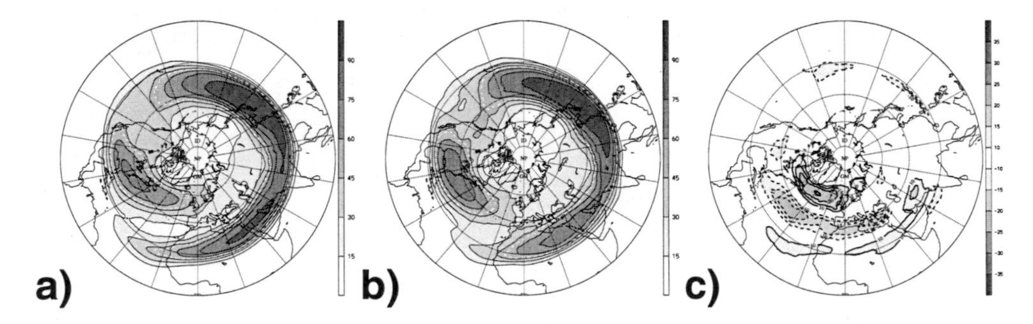

Figure 4.4: Winter jet event frequencies in % for (a) the NAO+ (mean over 28 months) and (b) the NAO- phases (mean over 17 months) and (c) differences in % between the NAO+ and NAO- jet event frequencies. Contours for (a) and (b) are analogous to Fig. 4.2. In (c), positive (negative) values are outlined in solid (dashed) and differences comprised between -10% and 10% are omitted.

Pacific-North America Pattern

The Pacific North American (PNA hereafter) teleconnection pattern is one of the most prominent modes of low-frequency climate variability in the Northern Hemisphere extratropics. It reflects a quadripole structure of height anomalies, with anomalies of similar sign south of the Aleutian Islands and over the southeastern United States and another two with the opposite sign in the vicinity of Hawaii, and over the intermountain region of North America[5]. In the positive phase, the Pacific storm track[6] and jet stream extend from East Asia over the central Pacific and into the Gulf of Alaska, while in the negative phase, the storm track and jet stream tend to turn directly northward from East Asia to the Bering Sea, with a small area of upper-level lows off the west coast of Canada (e.g. Barry

[5] e.g. http://www.cpc.ncep.noaa.gov/data/teledoc/pna.html

[6] i.e. here, upper-level cyclone trajectories

and Chorley 2003, Chapter 10).
The spiral shape of the patterns of jet event frequencies is well established in both PNA phases (Figs. 4.5 a and b) and is merdionally slightly wider in the eastern-Pacific sector in the negative phase. Jet events are much more common in the mid-Pacific in the PNA+ phase, as indicated by the strong positive signal in this region in Fig. 4.5 (c), with difference values exceeding 30%. This signal is part of a tripole structure of differences over the mid-Pacific/western-Pacific sector with two areas of negative frequencies near 20°N and over the Aleutian Islands. Strong signals are also found over North America, with a dipole structure composed of positive frequencies that extend from Mexico to the western Atlantic and negative frequencies over the United States. Note also a small region of positive frequencies over northern Canada. Besides the Pacific/Northern-America sector, only a few small variations of the jet event frequencies are found e.g. a small positive/negative dipole pattern south of Greenland, suggesting a slightly more northward turn of the eastern end of the North American pattern in the PNA+ phase. There are also slightly larger jet frequencies (10%) over the Canary Islands and western China and slightly smaller (10%) frequencies over the central Sahara and the Philippine Sea in the negative phase.

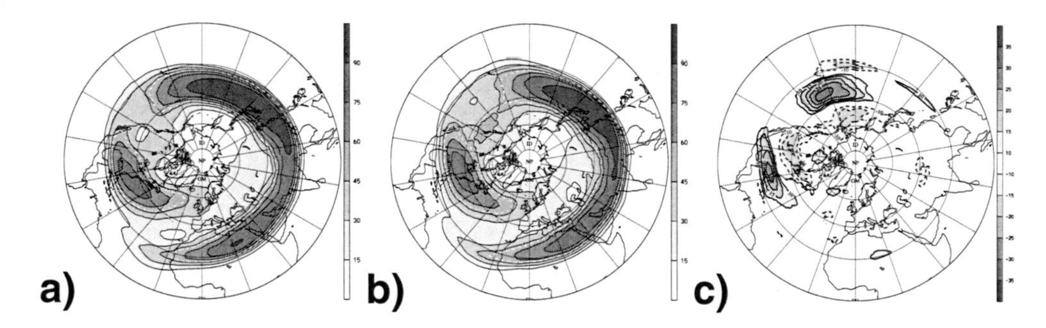

Figure 4.5: Same as Fig. 4.4 but for (a) the PNA+ (mean over 24 months) and (b) PNA- (mean over 18 months) phases.

Arctic Oscillation

The dominant EOF (Empirical Orthogonal Function) mode of the Northern Hemisphere variability (Thompson and Wallace 1998) defines the Arctic Oscillation (AO hereafter). The jet event patterns for the AO (Figs. 4.6 a and b) are similar to those obtained for the NAO (Figs. 4.4 a and b). The patterns of differences in the amplitude of the jet event frequencies (Fig. 4.6 c) show a tripole structure over the Atlantic with two positive signals extending respectively from the Great Lakes to Scandinavia and from the mid-Atlantic to the coasts of Mauritania. Between these two positive signals, a long zonal negative band extends from Mexico to Turkey. Over the Pacific, a positive difference is found between Japan and Alaska, and a meridionally narrow negative zone extends from Taiwan to northeast of Hawaii, this band being meridionally broader over the mid-Pacific. There is also a positive signal over the Red Sea, a dipole over the Arabian Sea (with the positive signal over Iran) and negative differences over Mongolia.

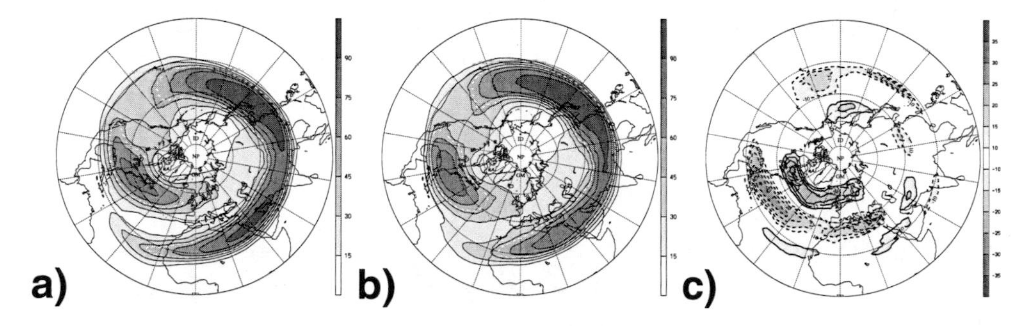

Figure 4.6: Same as Fig. 4.4 but for (a) the AO+ (mean over 22 months) and (b) AO- (mean over 23 months) phases.

4.3.2 Southern Hemisphere

Southern Oscillation

The Southern Oscillation (SO hereafter) is an integral part of the ENSO (El Niño Southern Oscillation). The cyclic warming and cooling of the eastern and central Pacific (characteristic of the El Niño/La Niña cycle) leaves its distinctive fingerprint on sea level pressure, e.g. the difference between the pressure measured at Darwin and that measured at Tahiti. When this difference is positive, there is a La Niña (or ocean cooling) phase and when negative there is an El Niño (or ocean warming) phase[7]. Moreover, the positive phase is accompanied by a strengthening of the low-level easterlies and a weakening of upper-level westerlies, whereas low-level westerlies dominate and upper-level westerlies tend to be stronger in the negative phase (Barry and Chorley 2003, Chapter 11).
Although the spiral pattern is evident in both phases (Figs. 4.7 a and b) it appears discontinuous over the Andes in the SO+ phase. The pattern of jet event frequency differences (Fig. 4.7 c) only shows the larger jet frequencies in the SO- phase in the form of a spiralling ribbon of negative amplitudes starting east of Australia and ending south of New Zealand.

[7]see also http://www.ogp.noaa.gov/enso/

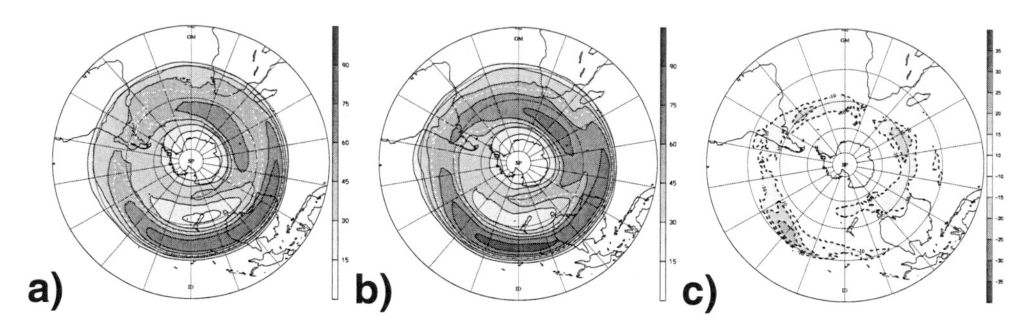

Figure 4.7: Same as Fig.4.4 but for (a) the SO+ (mean over 15 months) and (b) SO- (mean over 28 months) phases.

Antarctic Oscillation

The Antarctic Oscillation (AAO hereafter) is the southern counterpart of the boreal Arctic Oscillation. It also refers to a 'seesaw' variation of atmospheric mass between the mid- and high latitudes. Like the AO in the Northern Hemisphere, it is defined as the leading mode of the EOF analysis of the sea-level pressure (e.g. Gong and Wang 1999)[8].
The main difference between the AAO+ and AAO- phases of the spiral pattern of jet event frequencies (Figs. 4.8 a and b) is the vanishing of its poleward end south of New Zealand. This is clearly visible in the frequency differences (Fig. 4.8 c) in the form of a meridionally narrow band of positive amplitude located south of New Zealand. Note also the dipole structure of differences in the jet frequencies over the Indian Ocean, with the zone of positive differences poleward (upstream of the band described earlier) and the negative counterpart in the central Indian Ocean, suggesting a northward shift of the jet event pattern in the negative phase.

4.3.3 Discussion

The aim of this Section was to study the variation of the winter jet event climatology in relation to the major modes of variability of the global atmospheric circulation in both hemispheres, e.g. the impact of these flow variations upon the time and spatial occurrences of jet events is investigated. The indices used to quantify the variations of the global circulation are derived from lower-level variables and here the recurrent shifts in the positions of jets and the variations in amplitude of the occurrence frequencies are analyzed in relation to the mode

[8]However the CPC uses the monthly mean 700 hPa height poleward of 20°S, see http://www.cpc.ncep.noaa.gov/products/precip/CWlink/daily_ao_index/aao/aao_index.html.

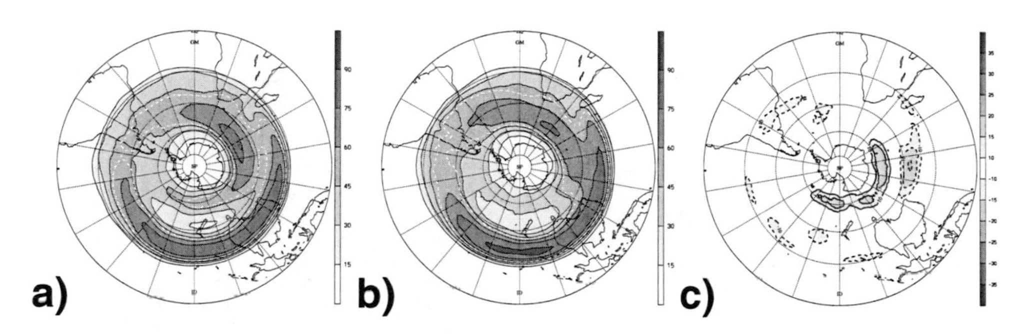

Figure 4.8: Same as Fig.4.4 but for (a) the AAO+ (mean over 24 months) and (b) AAO- (mean over 21 months) phases.

of flow variability. The results for the five variability modes obtained previously are summarized hereafter. Figure 4.9 illustrates the discussion. It depicts the variations of the 50% frequency patterns for both phases of the considered oscillation, and the pattern obtained directly from the winter jet event climatology of Section 4.2 for the whole ERA15 period.

Northern Hemisphere

NAO: Differences in the patterns of jet events frequencies are mostly located in the Atlantic/European sector. The eastward tail of the north American pattern extends farther downstream and tends to turn poleward over the western Atlantic in the positive phase while it is more zonally aligned and reaches only the mid-Atlantic in the negative phase. Furthermore, jet events are more frequently found over the Cape Verde Islands and western Mauritania in the positive phase, as shown by the westward extension of the upstream pattern present between Africa and the mid-Pacific.
PNA: Two main geographical regions are affected by variations from one phase to another: the eastern Pacific and the northern-America/western-Atlantic sectors. In the positive phase, jet events are more commonly found north and northeast of Hawaii, in the 20°N-40°N belt. The north American pattern tends to shift to the south in the positive phase, suggesting a general southward shift of the jet for this mode.
AO: The patterns for the AO share common points with those of the NAO and of the PNA. There is a poleward turn and an extension farther downstream of the north American pattern (a slight poleward shift of the whole pattern is also visible) and a development farther upstream of the pattern of jet event climatology over western Mauritania and the Cape Verde Islands during the positive phase. This suggests that, in the region comprised between 330°W and 0°, double-jet

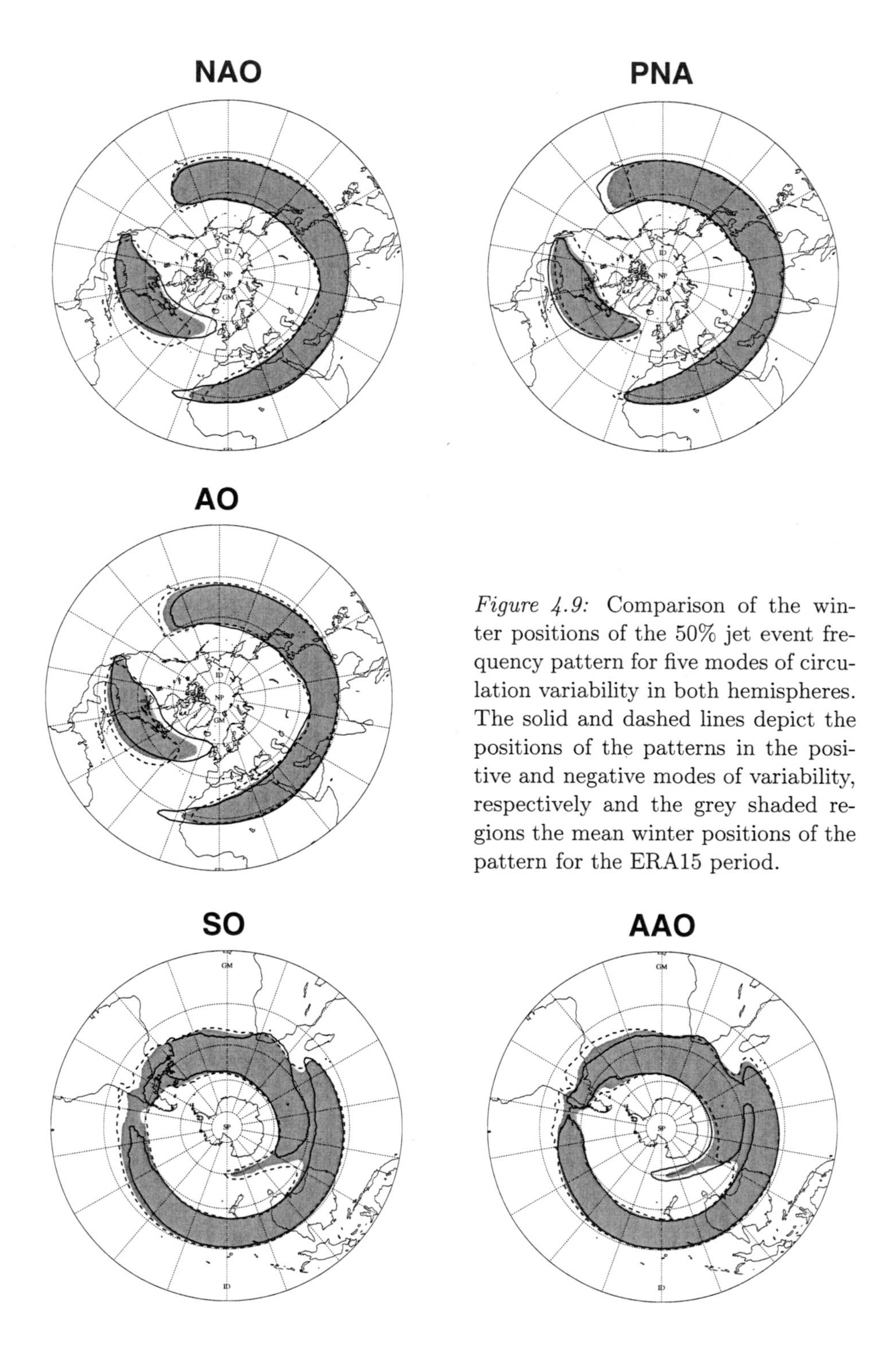

Figure 4.9: Comparison of the winter positions of the 50% jet event frequency pattern for five modes of circulation variability in both hemispheres. The solid and dashed lines depict the positions of the patterns in the positive and negative modes of variability, respectively and the grey shaded regions the mean winter positions of the pattern for the ERA15 period.

configurations are likely to occur during the positive phase. Moreover, another feature (not visible in the NAO) is present over the eastern Pacific. In the positive phase, jet events happen more rarely in the region northeast of Hawaii, which is also a feature of the negative phase of the PNA.

Southern Hemisphere

SO: No shift in the position of the jet event pattern is visible in the SO, as already suggested by the presence of only negative differences in Fig. 4.8 (c). It is actually the intrinsic structure of the pattern of jet frequencies that is modulated by the SO. An obvious feature of the positive phase is the drop in the frequency of jet event occurrences south of Australia (at nearly 40°S), together with the split into two distinct branches of the pattern over the Indian Ocean, suggesting a decrease of the number jet event occurrences slightly equatorward of 40°S for this phase. Note also a drastic decrease of jet event frequencies in the Andes region.
AAO: The negative phase of the AAO is characterized by the offset of the downstream tail of the spiral pattern of jet event frequencies in the midlatitudes, east of 120°E, showing a drop in the jet event occurrences in this region. The positive phase is characterized by a drop of the jet frequencies in the Andes region and in a narrow region slightly equatorward of 40°S over the mid- and eastern Indian Ocean, which suggests two preferred zonal bands for jet events in this region.

Further Remarks

It is noteworthy that variations in the Northern Hemisphere of the larger jet event pattern occur mainly at its upstream (Mauritania and eastern Atlantic, in the NAO and AO) and downstream (north and east of Hawaii, in the PNA and AO) tails. Note that the variations in this region are not the same for the positive phases of the PNA and AO (extension farther downstream in the PNA+ and AO- phases). No relevant shifts in the positions or the structure of the main body of the pattern is visible. The north American pattern exhibits a variation in its eastern tail as well as a meridional shift of its position in the PNA and AO (southward in the PNA+ phase and northward in the AO+ phase). In the Southern Hemisphere, no shift in the position of the spiral pattern of jet event frequencies is detected. The suppression of the poleward tail of the spiral is a characteristic feature of SO+ and AAO- phases. Note also the more seldom jet events in the Andes region and the separation of the two arms of the spiral over the Indian Ocean, owing to a decrease in the jet event frequencies in the SO+ and AAO+ phases.
Recently, an increased number of questions have arisen concerning the physical

relevance of the AO paradigm for the Northern Hemisphere in comparison to the NAO paradigm (Ambaum et al. 2001). Moreover, Deser (2000) showed that the teleconnectivity between the Arctic and the midlatitudes is strongest over the Atlantic sector and that the temporal coherence between the Atlantic and Pacific midlatitudes is weak, attributing more the annular character of AO to the dominance of its Arctic center of action. The results presented here enable a direct look at the differences at the tropopause level between the AO and NAO modes of variability. More precisely, the variations of regions and frequencies of jet event occurrences can be directly inferred. The AO and NAO jet event patterns are quite similar over the Atlantic sector, with a more poleward orientation and a further eastward extension of the north American jet event pattern in both positive phases. A distinctive feature appears over the Pacific where jet events are more common north and northeast of Hawaii in the AO- phase. No relevant difference is found in the jet occurrence over the Pacific in the NAO. The PNA features a clear variation in the distribution of jet events over the Pacific. However, the negative phases of the AO and PNA show opposite characteristics. In the PNA- phase, jet events are more difficult to find northeast of Hawaii. Noting this, the upper tropospheric jet structure associated with the positive phase of the AO appear to be comparable to the combination of the positive phase of the NAO and of the negative phase of the PNA and vice versa. Finally, using mean zonal wind studies, Hartmann et al. (2000) argued that the tropospheric jet is displaced poleward of its climatological position when the Northern Hemisphere annular mode is in its high index (i.e. AO+). The results of this study only agree with this statement when the northern-America/Atlantic sector is considered in which a slight northward shift of the north American pattern is visible. However no evidence of a meridional shift was found for the whole jet event pattern which extends from North Africa to the mid-Pacific.

4.4 Summary

In this Chapter a novel type of jet-stream climatology has been presented. It is based on the identification of jet events using an objective criterion based on the horizontal wind velocity in the tropopause region. Time and positions of these events are recorded and seasonal and monthly climatologies of the jet frequencies are calculated.

The seasonal jet event climatologies shed light on where and how often jet events occur while 'classical' jet-stream climatologies, based on the mean wind speeds on particular levels give an idea on how strong jets are on average at different levels.

The utility of the method has been illustrated with the study of the variations of the jet event frequency patterns in relation to the main variability modes of the global atmospheric circulation. This has allowed a direct insight into the variations of the time and space distributions of upper-level jet streams for flow variability modes, as defined by indices based on lower-tropospheric variation patterns. The main regions of variations have been defined. In the Northern Hemisphere, variations in the zonal length of the patterns are related to the phase of the variability mode. Variations in the meridional position of a jet pattern located over the northern- America/western-Atlantic sector have been identified. In the Southern Hemisphere, the largest variations are found equatorward of Australia and New Zealand, over the Indian Ocean and in the vicinity of Patagonia. Finally, differences between the AO and NAO jet event patterns have been briefly discussed.

Chapter 5

On the Baroclinic Characteristics of Jet Events

A study is undertaken of the vertical wind shear characteristics of the jet events. To this end, the vertical variations of the horizontal wind velocity are expressed by the difference of the wind speeds between the 200 and 500 hPa levels normalized by the wind velocity at 200 hPa. A partitioning of the jet events into two categories is then made based upon the values of this ratio. These two categories are an indicator of the vertical confinement of the jet core and of the associated baroclinic structure of the jet events and thereby permit a two-category classification of the jets.

5.1 Data and Methodology

The data used in this chapter are the jet events determined in Section 4.2 from the ERA15 data set. In order to quantify the vertical characteristics of the jet events, the decrease of the wind velocity between 200 and 500 hPa is calculated and normalized by the wind speed at 200 hPa:

$$\Delta v_{rel} = \frac{v_{200} - v_{500}}{v_{200}}, \tag{5.1}$$

where v_{200} and v_{500} are the horizontal wind velocities at 200 and 500 hPa respectively. The reason for the choice of the 200 hPa and 500 hPa levels is related to the process of distinguishing the subtropical jet from the polar-front jet. The subtropical jet core is usually found around 200 hPa and its associated meridional temperature gradient and vertical wind shear are concentrated in a shallow layer (unlike the polar-front jet, e.g. Chapter 3 and Bluestein 1986 Section 2.7).

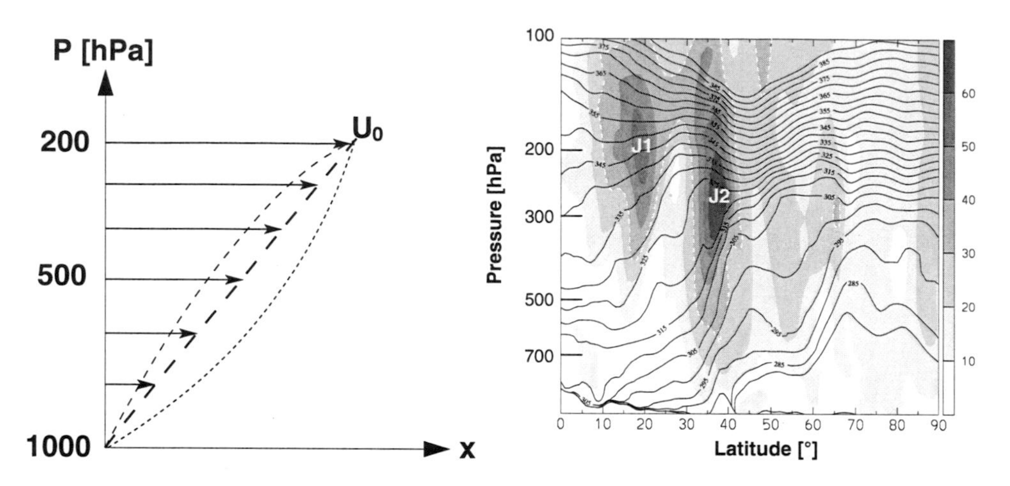

Figure 5.1: Schematic of a simple vertical linear wind profile (left panel). The dotted (short-dashed) curve sketches a concave (convex) profile. The vertical coordinate is pressure in hPa. Vertical cross section of wind velocity (right panel, in ms^{-1}, white dashed line outlining the 30 ms^{-1} isopleth) and potential temperature (solid lines, contours every 5 K from 290 K to 380 K) at 30°E for 00 UTC, 3 January 1979. J1 and J2 denote the two jets discussed in text.

Thus, by comparing the wind speed at 200 hPa and in the middle troposphere (500 hPa), and by calculating the ratio Δv_{rel}, a direct and compact measure of the vertical wind shear is gained.
The choice of a threshold for the subsequent partitioning of the jet events into two categories is based upon considerations made on a simple vertical wind profile similar to the one proposed in the Eady model (Eady 1949), except for the vertical coordinate (pressure instead of height). The left panel of Fig. 5.1 shows such a vertical wind profile. The values of the wind velocity at the surface and at 200 hPa are respectively 0 and U_0. It can be shown that for this configuration, the value obtained for Δv_{rel} is 0.375. The study of other vertical wind profiles show that concave (convex) profiles are associated with smaller (larger) values of Δv_{rel}. In this study, the threshold criterion for Δv_{rel} used in the classification of the jet events is set to 0.4.
The first category gathers jet events having a Δv_{rel} larger than the threshold and featuring a vertical wind shear concentrated in the upper troposphere. It will be designated as the shallow layer baroclinicity category (SL category hereafter). The jet events of the second category have a Δv_{rel} smaller than the threshold and are characterized by a vertical wind shear that can extend much further down beneath the upper troposphere into the lower troposphere. This category will be referred to as the deep layer baroclinicity category (DL category hereafter). By adding the jet events of the SL and DL categories, the seasonal jet event climatologies are retrieved.

As an illustrative example, the right panel of Fig. 5.1 shows a vertical cross section at 30°E of the horizontal wind velocity and potential temperature for the same time instance as Fig. 4.1. The evaluated values of Δv_{rel} for the two jets J1 and J2 are 0.8 and 0.3, categorizing them as members of the SL and DL categories, respectively. Note also the differences in the baroclinic characteristics of the two jets: J1 is associated with a baroclinic zone that extends down to only about 500 hPa, while J2 has a baroclinic zone extending down to the surface.

5.2 Seasonal Climatologies

In this section, the interseasonal variations of the jet event patterns are presented and discussed for the SL and DL categories in both hemispheres.

5.2.1 Northern Hemisphere

SL Category

Winter (Fig. 5.2 DJF) exhibits a spiral like pattern of jet events in the 15°N-40°N belt with one end over the central Atlantic west of the Cape Verde Islands and the other some 20° poleward. Jet events are commonly found between Africa and the mid-Pacific (with frequencies exceeding 75% and 90% over Egypt and eastern China, respectively) and over the southern coasts of the United States (with frequencies that exceed 50%).
Spring (Fig. 5.2 MAM) is characterized by an irregular ring pattern confined mostly inside the 20°N-40°N belt with a meridional discontinuity found in the Hawaiian sector. Note that the meridional gap is no longer visible over the Atlantic as was the case in winter. Three regions are visible where jet event occurrences are common: a first one over North Africa and the Arabian Peninsula (with frequencies not exceeding 60%), a second one over eastern Asia and Japan (with frequencies exceeding 60%) and a third one extending from east of Hawaii to the Bahamas (with frequencies less than 50%).
In summer (Fig. 5.2 JJA), the spring ring-shaped jet event pattern shrinks to become two arcs (one extending from Turkey to the mid-Pacific and the second one extending from the Great Lakes to Nova Scotia). Note also a poleward migration of some 15° of the summer patterns, in comparison to spring and the passage of the pattern to the north of the Himalaya. Only a very few jet event occurrences are found in summer as indicated by the very low frequencies.
Autumn (Fig. 5.2 SON) presents a more developed version of the summer jet event pattern with an open spiral shape from the Sahara to the mid-Atlantic and

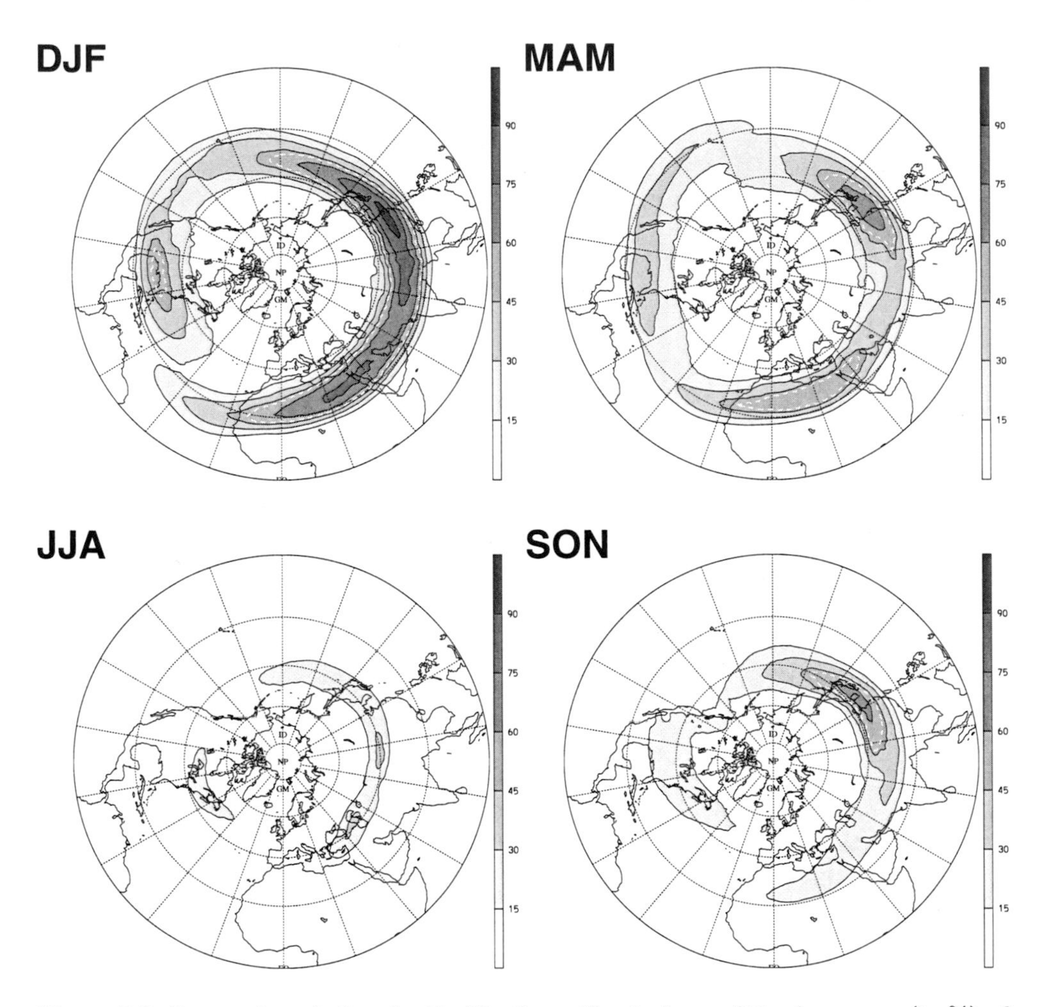

Figure 5.2: Seasonal evolution for the Northern Hemisphere of the frequency (in %) of the jet events of the SL category for a threshold value of 0.4. The dashed white line indicates 50%. Seasons are winter (DJF), spring (MAM), summer (JJA) and autumn (SON).

an irregular equatorward extension of the frequency pattern, ranging from 5° to 15° in some regions. There is a large increase of the jet event frequencies over eastern Asia and the western Pacific, with values exceeding 60% over Korea and Japan.

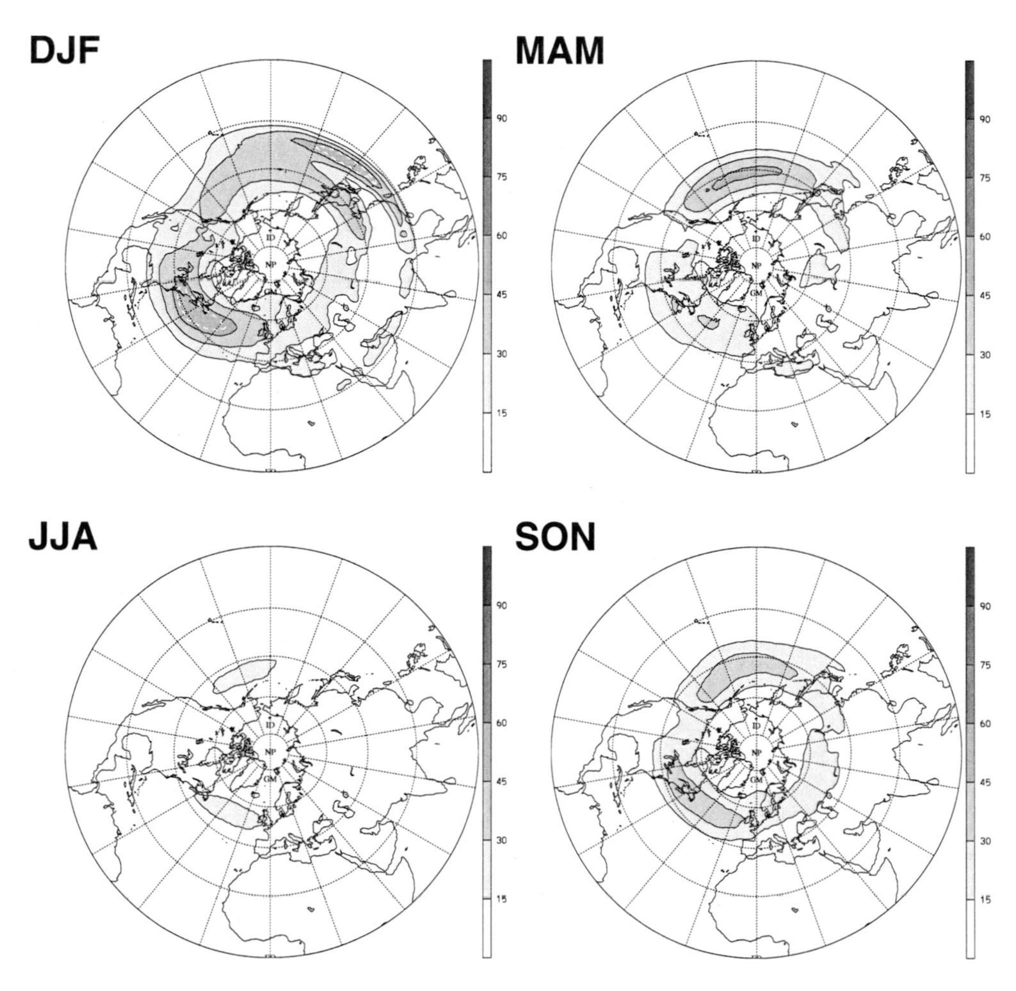

Figure 5.3: Same as Fig. 5.2 but for the DL category. Seasons are winter (DJF), spring (MAM), summer (JJA) and autumn (SON).

DL Category

The winter pattern of jet event frequencies (Fig. 5.3 DJF) exhibits an irregular spiral shape with a disrupted meridionally narrow band, extending from the Arabian Peninsula to the mid-Pacific, and a distorted ring-like pattern mainly poleward of 30°N. Jet frequencies exceeding 50% are found over the western Pacific (between 20°N and 30°N) and over the western and mid-Atlantic north of 40°N.
In spring, the winter jet event pattern (Fig. 5.3 MAM) takes an open ring shape extending from eastern Asia to the eastern Atlantic, disrupted in the region of the Rocky Mountains. The areas where jet events are mostly found are located

over the mid-Pacific and the mid-Atlantic.
No coherent pattern of jet events is visible in summer (Fig. 5.3 JJA), except for two small regions of very low frequencies located poleward of 40°N over the Pacific and the Atlantic.
Autumn (Fig. 5.3 SON) has a ring-like jet event pattern with a geographic location similar to that in spring. Jet event occurrences are primarily detected poleward of 40°N over the Pacific and from the Great Lakes to the Atlantic.

5.2.2 Southern Hemisphere

SL Category

The winter jet event pattern (Fig. 5.4 JJA) has a continuous ring-shape in the 20°S-40°S belt. Jet events are more commonly found from the western Indian Ocean, south of Mauritius, to the mid-Pacific with maximum frequencies over Australia and northwest of New Zealand (values exceeding 90%).
Spring (Fig. 5.4 SON) exhibits an irregular ring pattern confined in the 20°S-40°S belt, except from the Atlantic to the Indian Ocean, where a branch of low frequencies extends southward to 60°S. Jet events occur mainly from the west of Australia to the western Pacific sector, with peaks in the frequencies exceeding 50% over western Australia and the western Pacific sector.
In summer (Fig. 5.4 DJF), only two arcs of low frequency remain from the spring pattern. One extends from South America to south of Madagascar and the second from southwestern Australia to the mid-Pacific sector. Note that no relevant poleward migration of the jet event pattern is visible here in comparison to spring.
The jet event pattern for autumn (Fig. 5.4 MAM) is a disrupted ring with a zonal gap south of Madagascar. Jet events are more common from the west of Australia to the mid-Pacific with a peak in the frequencies (over 40%) southeast of New Caledonia.

DL Category

The winter jet event climatology (Fig. 5.5 JJA) exhibits a spiral pattern with a branch extending from Australia to the eastern Pacific sector where it merges with a ring pattern of jet frequencies. Jet events are more commonly found in this ring, poleward of 40°S, from the mid-Atlantic to the southeastern Indian Ocean, with local maximum values exceeding 60% in the Indian Ocean.
In spring (Fig. 5.5 SON), a ring jet event pattern emerges, circling Antarctica, on average poleward of 40°S. The winter spiral-like pattern is barely distinguishable. The mid-Atlantic/Indian Ocean sector is the region where jet events are most

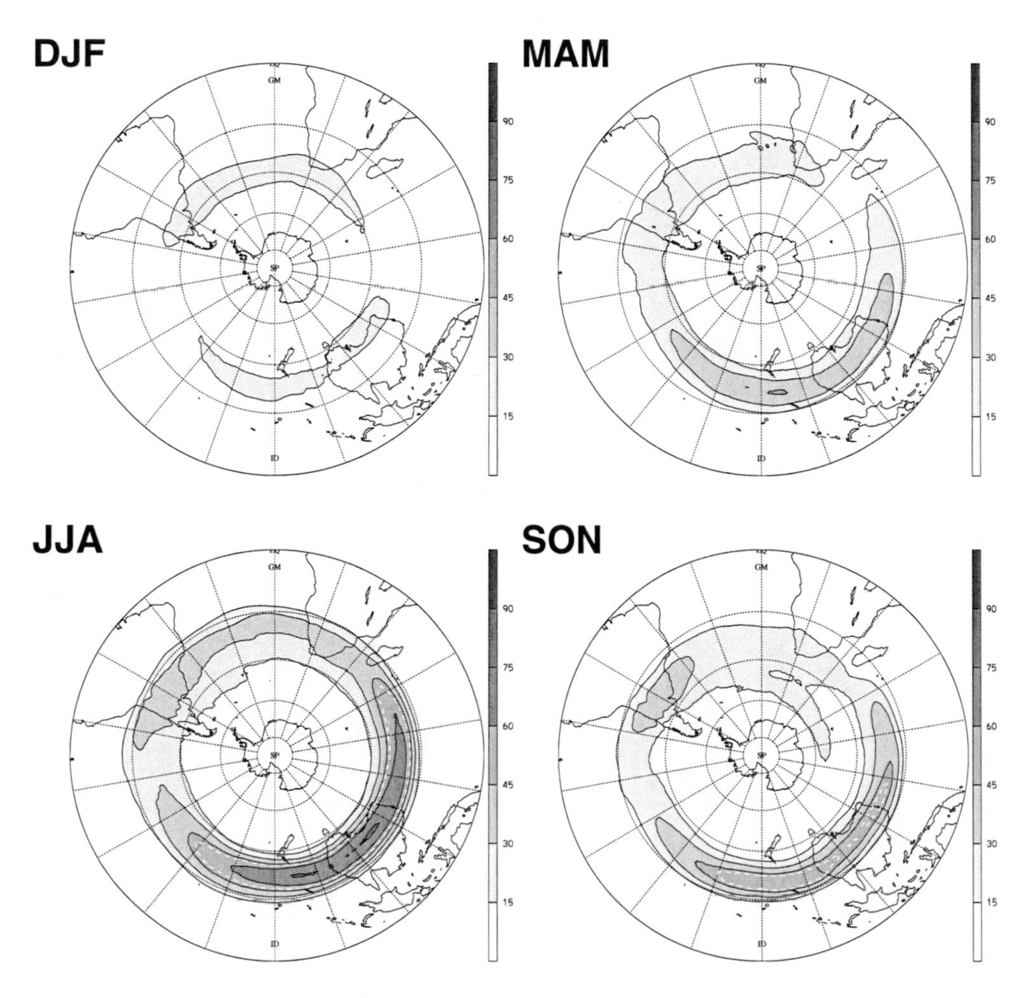

Figure 5.4: Seasonal evolution for the Southern Hemisphere of the frequency (in %) of the jet events of the SL category for a threshold value of 0.4. The dashed white line indicates 50%. Seasons are summer (DJF), autumn (MAM), winter (JJA) and spring (SON).

commonly found with a peak in the frequencies of over 60% in the Indian Ocean. In summer (Fig. 5.5 DJF), a well established ring-like jet event pattern is found that surrounds Antarctica in the 40°S-60°S belt. Regions with the highest rate of jet event occurrence are, like in spring, concentrated from the mid-Atlantic to the southeastern Indian Ocean.
Finally, autumn (Fig. 5.5 MAM) displays a slightly meridionally broader version of the summer ring-like jet event pattern. The region where jet events are the more commonly found extends from the mid-Atlantic to south of New Zealand.

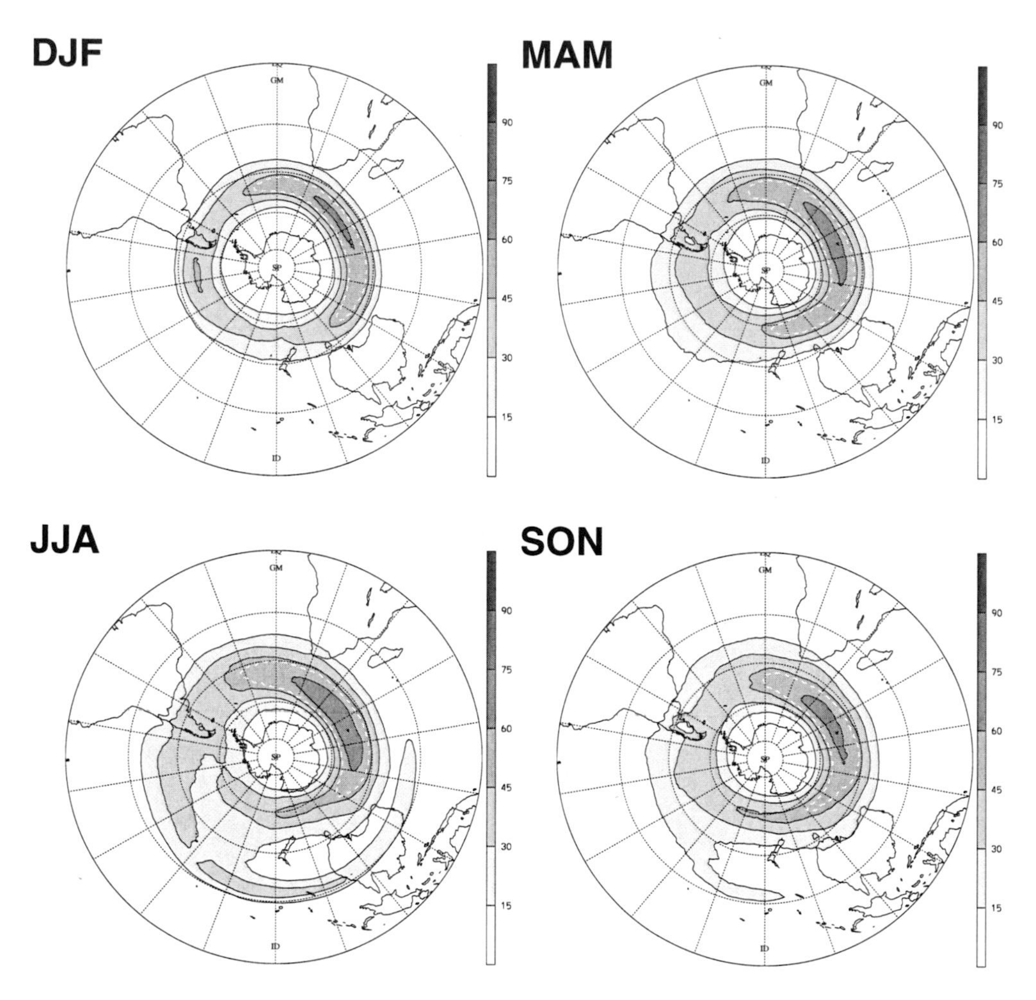

Figure 5.5: Same as Fig. 5.4 but for the DL category. Seasons are summer (DJF), autumn (MAM), winter (JJA) and spring (SON).

5.2.3 Discussion

General characteristics of jet event patterns and amplitudes in each hemisphere can be inferred from this study. Winter SL and DL jet event frequency patterns for different thresholds of Δv_{rel} have been calculated for both hemispheres as a further illustration of the influence of the choice of the threshold on the distributions of jet events into the SL and DL categories and can be found in the Appendix (see Figs. B.2 and B.3).
The seasonal variability of each category is not the same for each hemisphere. In the Northern Hemisphere, the variability is quite high for both categories. The pattern of SL jet events evolves from a spiral in the 20°N-40°N belt in winter,

with locally frequencies exceeding 90%, to two small arcs of very low frequencies located farther poleward (north of 35°N). In an analogous manner, the jet event pattern of the DL category evolves from a highly distorted spiral, with local maxima larger than 50% over the western Pacific and the Atlantic, to two very weak signals in summer, in the Pacific and the Atlantic, north of 40°N. On the contrary, the seasonal variations in the Southern Hemisphere are less relevant for the SL category and nearly inexistant for the DL category. The pattern of the SL category evolves from a ring in the 20°S-40°S belt, with maximum frequencies exceeding 90%, to two arcs of low frequencies with no major poleward shift detected. Note also the quasi-axisymmetry of the SL category pattern that is maintained throughout the year. The DL jet event pattern is characterized throughout the year by an annulus in the 40°S-60°S belt, evolving into a spiral-like feature in winter. Note that jet events occur most of the time (frequencies larger than 50%) from the mid-Atlantic to the southeastern Indian Ocean throughout the year.

5.3 Subtropical and Polar-Front Jets

The two-category partition of the jet events is based upon considerations of the vertical wind shear characteristics. The climatologies presented in Section 5.2 have shown that shallow-layer baroclinic (SL) jet events are found equatorward of 40° in winter and farther poleward in summer (Figs. 5.2 and 5.4). Deep-layer baroclinic (DL) jet events are generally found poleward of 30°, except in winter where distinct frequency patterns are also found poleward of 20° (Figs. 5.3 and 5.5). In the literature, a distinction is made between two types of upper-level jet streams with characteristic properties (see Chapter 3): the subtropical jet (STJ) and the polar-front jet (PFJ). The aims of this Section is to characterize the STJ and PFJ in the perspective of the SL and DL categories and to propose the two categories as a novel alternative classification.

5.3.1 Subtropical Jets and the SL Category

The jet events of the SL category are characterized by a large vertical wind shear in the upper troposphere (expressed here as a ratio of the wind speed difference between the 200 hPa and 500 hPa levels to the wind speed at 200 hPa larger than 0.4). As the decrease of the wind speed is large in the upper troposphere, the associated baroclinicity is also expected to be concentrated in the upper troposphere. Furthermore, the winter climatologies presented in Section 5.2 have shown that jet events of the SL category tend to be found in the 20°-40° zonal belt. Some common characteristics seem to exist between SL jet events and STJ. STJ are characterized by a baroclinic zone concentrated in a shallow layer at upper-levels (e.g. Table 3.1). However, discripancies between SL jet events and the definitions of STJ arise. First, the STJ is defined as a typical wintertime phenomenon while SL jet events are found throughout the year in both hemispheres. Second, the STJ is expected to be found in the 20°-35° zonal belt (e.g. Table 3.1, Krishnamurti 1961), while jet event frequency patterns are found from 15° to 45°. Finally, Bluestein (1993) states that in the Northern Hemisphere summer only the PFJ appears in mean wind speed charts at 200 hPa, while SL jet events are found in summer around 40°N in Fig. 5.2, which is consistent with the summer poleward displacement of the Hadley Cell. It appears then that the definition of the SL jet events contains the definition of the STJ and is based on a physically consistent property of jet events (vertical wind shear).

5.3.2 Polar-Front Jets and the DL Category

The jet events of the DL category are characterized by a relatively small vertical wind shear in the upper troposphere (ratio smaller than 0.4). This implies that the associated baroclinicity extends over a deep layer of the troposphere. The climatologies of Section 5.2 show that DL jet events are found throughout the year in both hemispheres. In general they are located poleward of 35° in both hemispheres, except in winter, where patterns of jet event frequencies are found poleward of 20° (e.g. Figs. 5.3 and 5.5). This latter fact is in contradiction with some of the definitions of the PFJ given in Table 3.1 (position of the PFJ and link to the midlatitude polar front). The definition of the DL jet events appears to contain the definitions of the PFJ.

5.3.3 A Novel Two-Type Classification for Jet Streams

The use of the definitions of Table 3.1 for the distinction between the STJ and PFJ is neither practical nor rigorous and the identification of the STJ or PFJ on weather charts is a difficult task. It is suggested here that the distinction between SL and DL jet events can be used as an alternative and more complete definition in order to distinguish between two different types of upper-level jet streams formerly identified by the terms 'subtropical' and 'polar-front'.
In Sections 5.3.1 and 5.3.2, it has been found that, although based on similar baroclinic properties, the SL and DL jet events were not exactly equivalent to the STJ and PFJ, respectively (following definitions found in textbooks). STJ and PFJ appear to actually belong to the SL and DL categories, respectively. This new categorization is directly based on the vertical wind shear characteristics of jets (that are similar to the characteristics of STJ and PFJ). Figure 5.6 shows the 20% frequency patterns of the SL (grey shaded) and DL (hatched)

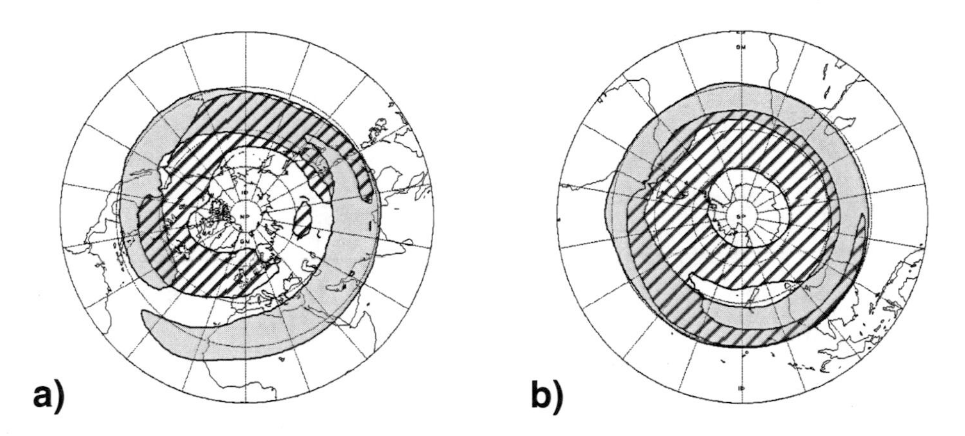

Figure 5.6: Geographical distribution of the 20% frequency patterns of jet events of the SL (grey shaded) and DL (hatched) categories for (a) the Northern and (b) the Southern Hemisphere winters.

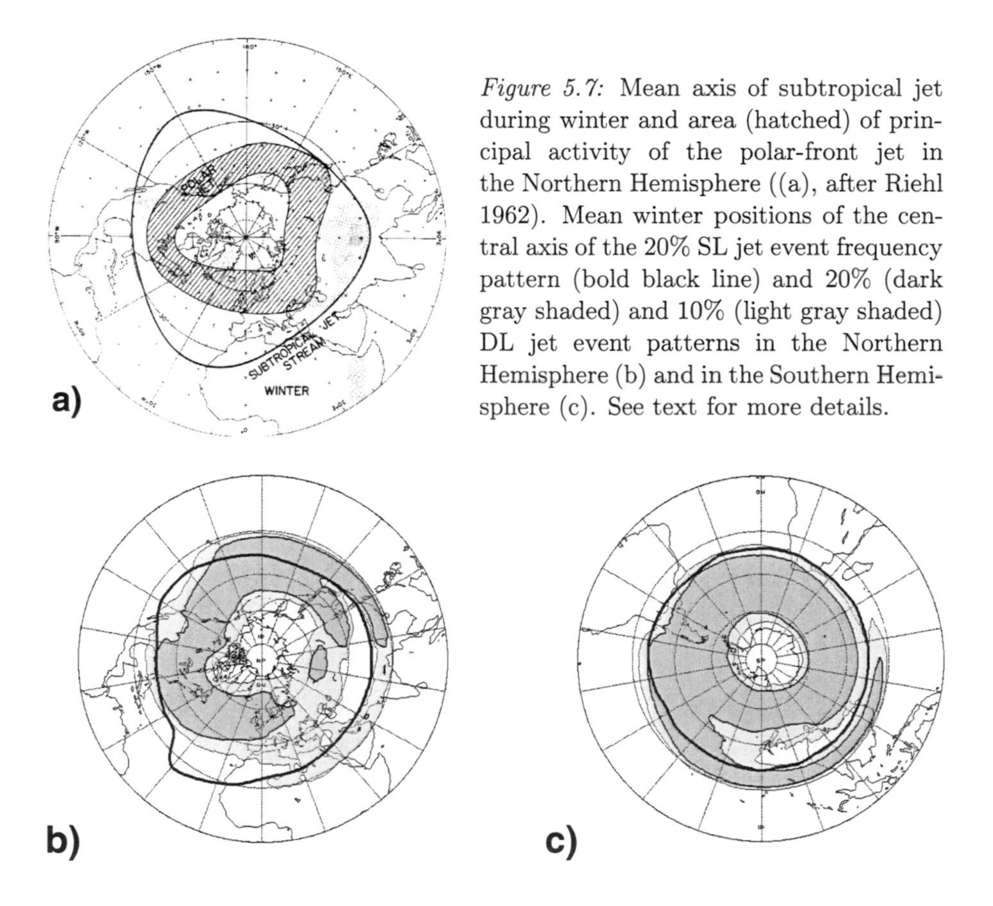

Figure 5.7: Mean axis of subtropical jet during winter and area (hatched) of principal activity of the polar-front jet in the Northern Hemisphere ((a), after Riehl 1962). Mean winter positions of the central axis of the 20% SL jet event frequency pattern (bold black line) and 20% (dark gray shaded) and 10% (light gray shaded) DL jet event patterns in the Northern Hemisphere (b) and in the Southern Hemisphere (c). See text for more details.

categories for both hemispheres.

Winter in the Southern Hemisphere displays an axisymmetric distribution of the frequency patterns of each category, these latter being mostly concentric around the Pole (Fig. 5.6 b). Except for the branch extending from Australia south of 20°S to South America some 20° poleward, the pattern of DL jet events does not extend equatorward of 35°S. The region of low frequencies already discussed in Chapter 4 extends from Tasmania to the western Pacific and appears here to be a region between the SL and DL jet event patterns. Regions exist where the SL and DL patterns overlay each other (from the eastern Indian Ocean to the eastern Southern Pacific in the 20°S-40°S belt and around 40°S from South America to southwest of Australia). This implies that in these regions SL and DL jet events can be found.

On the contrary, the boreal winter (Fig. 5.6 a) offers a lesser axisymmetric pattern. First, the SL jet event pattern has a disrupted ring shape, as a discontinuity is present over the mid-Atlantic with a meridional gap of about 10° to 15°. Regions with overlaid patterns of SL and DL jet events are found from eastern Asia to the mid-Pacific and from the Middle-West to the western Atlantic. Note that there is a split of the mean winter upper-level flow in the mid-Atlantic sector (not shown) with a branch turning northeastward and accelerating into the DL pattern, and a second one turning southeastward, decelerating before re-accelerating in the region of the SL pattern. This configuration leads to the meridional separation of both patterns and suggests conditions that might favour the occurrence of double-jet events (e.g. Chapter 6). Finally, there is a double-branch pattern of the DL jet event frequencies extending from the east of the Himalayas to the western Pacific sector. The absence of DL jet events over central Asia illustrates the role played by the Himalayas in the separation of airmasses in the lower- and middle-troposphere, which in turn has a direct impact on the distribution of baroclinic zones in the lower- and middle-troposphere.

Riehl (1962) proposed a schematic of the mean positions of the axis of the subtropical jet and of the area of activity of the polar-front jet (Fig. 5.7 a). It depicts regions where both features are meridionally close (i.e. the western Pacific, eastern United States and the Near East), while other regions exhibit well separated features (i.e. the Himalayan sector, the eastern Pacific and the eastern Atlantic/European sector). A novel version of this schematic is proposed in Fig. 5.7 (b), as well as the analogous for the Southern Hemisphere in Fig. 5.7 (c), using SL and DL jet event frequency patterns. They portray the mean winter position of the central axis of the 20% SL jet event frequency pattern and the 20% (dark gray) and 10% (light gray) DL jet event frequency patterns.

The mean SL pattern axis found in Fig. 5.7 (b) has no marked wavenumber 3 pattern as the STJ axis in Riehl's schematic. It is shifted slightly southward over the eastern Pacific and farther more over the mid-Atlantic, but is mainly zonal in the Himalayan sector. The 20% DL pattern does not circle the hemisphere (discontinuity over Siberia) and has a spiral-like pattern with a branch extending

from southeastern China to the mid-Pacific. Note that for the 10% DL pattern, the inner-ring is continuous over Siberia and the outer branch extends farther westward to Libya, where it connects to the inner-ring pattern. The whole pattern takes the form of a compact spiral with regions without signal extending from Central Asia to the Himalaya, in contrast to Riehl (1962). The 10% DL pattern exhibits a split into two branches over the Caspian Sea. This suggests the possible occurrence of jet events of the DL category from southern Libya to the south of the Himalaya.
Globally, the Southern Hemisphere presents more zonal patterns than the Northern Hemisphere. However, the 10% DL patterns are similar in both hemispheres (a spiral pattern) but are more perturbed in the Northern Hemisphere, illustrating the forcing effects of the land masses and orography on the distribution of jets with shallow layer and deep layer baroclinic characteristics.

5.4 DL Jet Events and Storm Tracks

The fact that upper-level jet streams are connected to baroclinic zones also implies that there is a relationship between tracks of transient synoptic systems (e.g. storm tracks) und upper-level jets. Figures 5.8 (b) and (d) show winter climatological frequencies of the positions of moving surface cyclones in the Northern and Southern Hemispheres, respectively, determined from the ERA15 data set using the method developed by Wernli and Schwierz (2004) for the ERA40 data set[1]. The comparison of these figures with the winter climatological jet frequencies of Chapter 4 (Figs. 4.2 a and c) raises issues on how to relate these two kinds of climatologies. However, as already stated earlier, the baroclinic zones associated with upper-level jet streams can either be shallow and confined to the upper troposphere or extend over a deep layer down to the middle- and lower troposphere. This latter category is the most relevant for transient synoptic systems at lower levels. It is then instructive to compare Figs. 5.8 (b) and (d) with the winter jet event patterns of the DL category (reproduced in Figs. 5.8 a and c).
In the Northern Hemisphere, the maximum frequencies of the cyclone climatology are located in the midlatitudes over the Pacific (between Kamtchatka and the Aleutian Islands) and over the Atlantic (between eastern Canada and Scandinavia). These maxima are mainly located on the poleward side of the DL jet event patterns (e.g. the cyclonic shear side). Note that no feature equivalent to the spiral branch, extending from the Arabian Peninsula to eastern China, is found in the cyclone climatology.
In the Southern Hemisphere, the maxima of the cyclone climatology are found

[1]In their study, Wernli and Schwierz (2004) identify cyclones as the largest possible region bounded by a closed contour of constant sea-level pressure, in a way that this contour encloses only one local sea-level pressure minimum

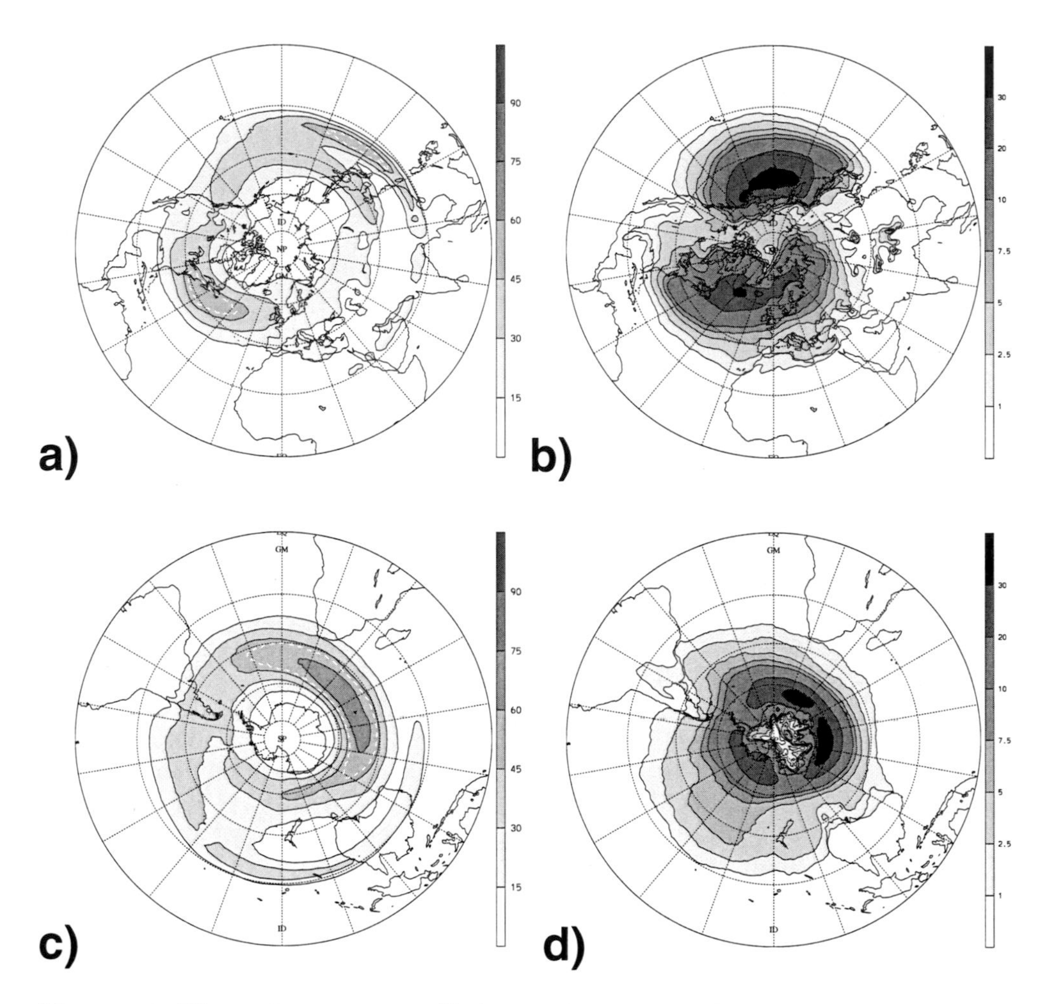

Figure 5.8: Winter frequencies in % derived from the ERA15 reanalysis data set of the DL jet events for (a) the Northern Hemisphere (same as Fig. 5.3 a) and for (c) the Southern Hemisphere (same as Fig. 5.5 c). Winter frequencies of moving surface cyclones from Wernli and Schwierz (2004) in % for (b) the Northern Hemisphere and for (d) the Southern Hemisphere.

in the southern Indian Ocean and southern Pacific. Like in the Northern Hemisphere, these maxima lie on the poleward side of the DL jet event pattern. Note that the region of enhanced cyclone frequency east of New Zealand is located on the poleward side of the spiral branch that extends from Australia to the mid-Pacific. An enhancement of the cyclone frequencies is found over the western Southern Pacific, equatorward of 40°S and poleward of the arm of the spiral pattern of DL jet events that extends from Australia to the mid-Pacific.
This analysis shows that when the jet event climatology is partitioned into the

SL and DL categories, there is no correspondence between the SL category and storm tracks. It also illustrates the clear correspondence between DL jet events and transient surface weather systems in the midlatitudes over oceans in both hemispheres. The seasonal evolution of both the DL jet event patterns and cyclone climatologies shows the same general correspondence in the midlatitudes. This is further illustrated in Figs. B.9 and B.10 in the Appendix.

5.5 Interannual Variations

The separation of jet events into the SL and DL categories enables a further refinement of the analysis of Section 4.3. The aim of the present section is to determine to which extend the winter repartition of SL and DL jet events vary with the major modes of global flow circulation variability. This can in turn give information on the baroclinic state of the atmosphere in each mode of variability, especially in regions where jet events of both categories are very close.

5.5.1 Northern Hemisphere

North Atlantic Oscillation

The features forming the tripole found over the Atlantic in Fig. 4.4 (c) appear to be distributed among the SL and DL categories. The positive signal at around 20°N (Fig. 5.9) refers to an increase of SL jet events in the positive phase. The dipole signal south of Greenland (and the associated northward shift of the eastern tail of the north American pattern in the positive phase) are clearly members of the DL category. Note also that the negative signal over the Mediterranean in Fig. 4.4 c. might be attributable to a slight equatorward shift of the SL pattern in this region in the positive phase.

Pacific-North America Pattern

The large signal in the differences of jet event frequencies over the Pacific in Fig. 4.5 (c) can be decomposed into two contributions. First, the SL jet event pattern turns slightly more northward in the central Pacific and extends a little farther east of 160°W in the positive phase (Fig. 5.10). Second, there is a southern shift of the DL pattern in the mid-Pacific in the positive phase, together with a slight downstream elongation east of 160°W. Note also that the southward shift of the north American pattern detected in Fig. 4.5 is mainly due to a shift of SL jet events.

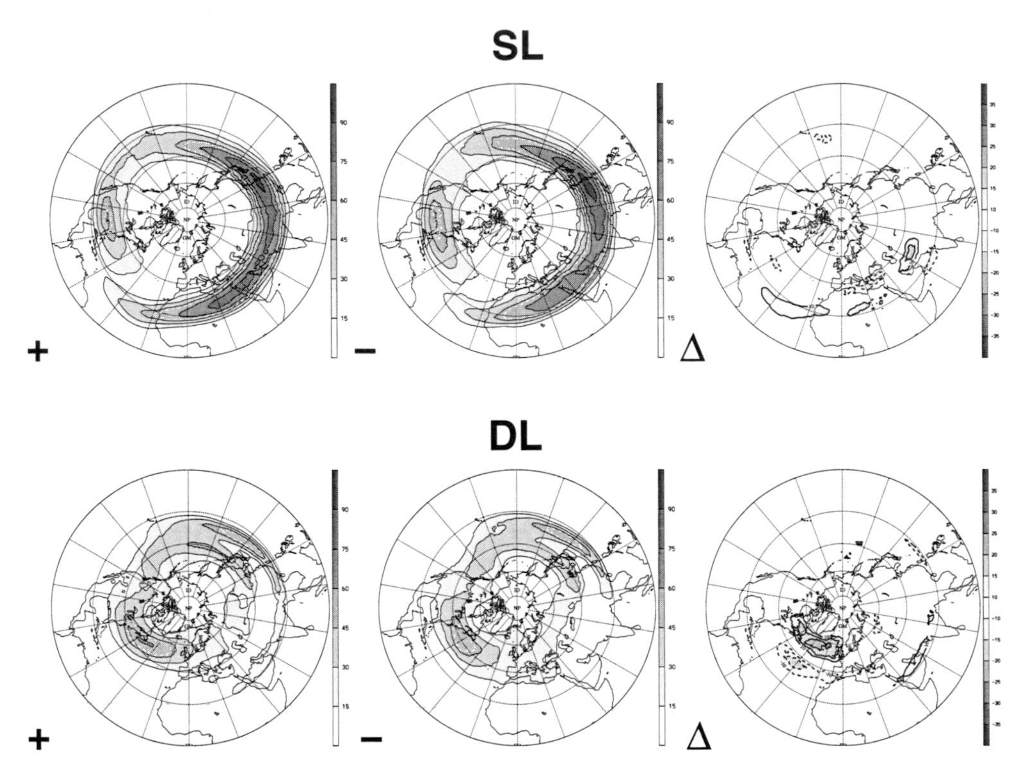

Figure 5.9: Winter jet event frequencies (in %) for the SL (upper row) and DL categories (lower row) for the NAO. The + and - indicate the positive and negative phases and Δ the difference between the amplitudes of the patterns of the positive and negative phases. Contours are the same as Fig. 4.4.

Arctic Oscillation

As for the NAO, the positive difference found west of Africa at nearly 20°N (Fig. 5.11) denotes an increase in SL jet events in the positive phase in this region. Although the long zonal positive difference between the Great Lakes and Scandinavia is clearly a DL feature, both categories contribute to the negative signal extending from the Gulf of Mexico to Turkey (e.g. Fig. 4.6 c). The western end of the negative signal (from the Gulf of Mexico to the mid-Atlantic) is a contribution from SL jet events, while the portion over the mid- and eastern Atlantic is due to jet events of the DL category. Finally, the dipole visible in the mid-Pacific in Fig. 4.6 (c) is actually due to (i) a downstream extension of the SL pattern, in the negative phase of the AO, for its southern component and (ii) to an increase in the occurrence of DL jet events in the Bering Sea, in the positive phase of the AO, for its northern component.

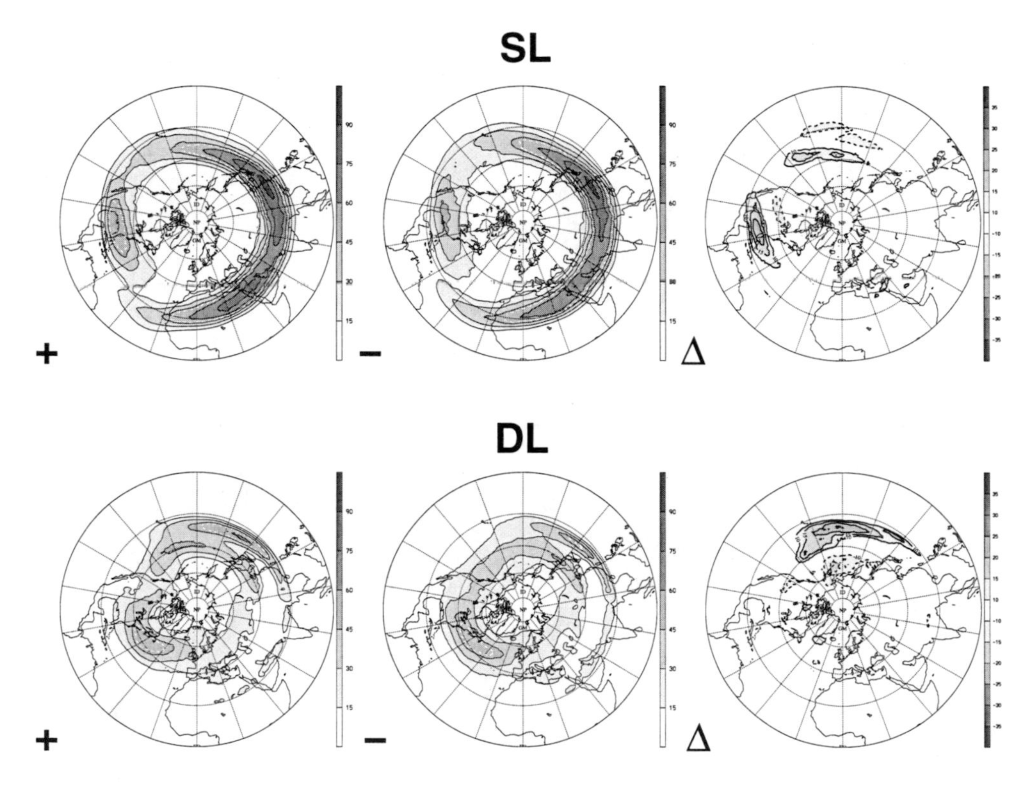

Figure 5.10: Same as Fig. 5.9, but for the PNA.

5.5.2 Southern Hemisphere

Southern Oscillation

The negative phase of the SO corresponds to an increase of SL and DL jet event occurrences over the western and mid-Pacific sector and to an increase of jet events of the DL category poleward of 40°S from the Indian Ocean to the south of New Zealand.

Antarctic Oscillation

The negative phase of the AAO corresponds to small regions of increased SL jet event occurrences distributed in the 20°S-40°S (Fig. 5.13). However, the positive phase corresponds to a poleward shift of jet events of the DL category in the Indian Ocean and to an increase of their frequencies from Tasmania to the southeast of New Zealand.

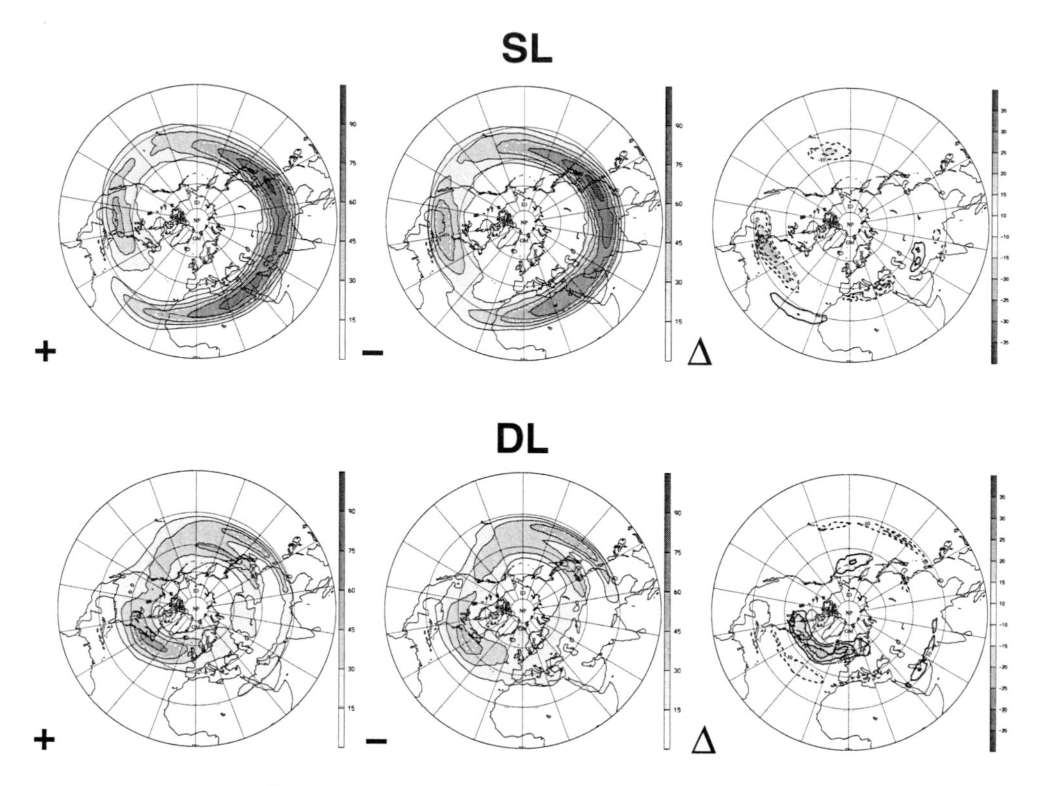

Figure 5.11: Same as Fig. 5.9, but for the AO.

5.5.3 Discussion

The analysis of interannual variations of jet events of the SL and DL categories gives new insight into the major variability modes of the global atmospheric circulation by studying the structure and amplitude of upper tropospheric jet event frequency patterns of different baroclinic properties. This distinction is particularly interesting in the Northern Hemisphere, where regions with very similar patterns for SL and DL jet events are found.
Some variability modes are manifest for one category of jet events. Thus, the NAO and the AO show significant variations in the DL jet event patterns. The other modes of variability show significant variations in both jet categories.
This analysis also gives some further distinctions of the characteristics of the NAO, PNA and AO by comparing Figs. 5.9, 5.10 and 5.11. In Section 4.3, it was suggested that the positive AO phase could share common features with the positive phase of the NAO and negative phase of the PNA. The comparison

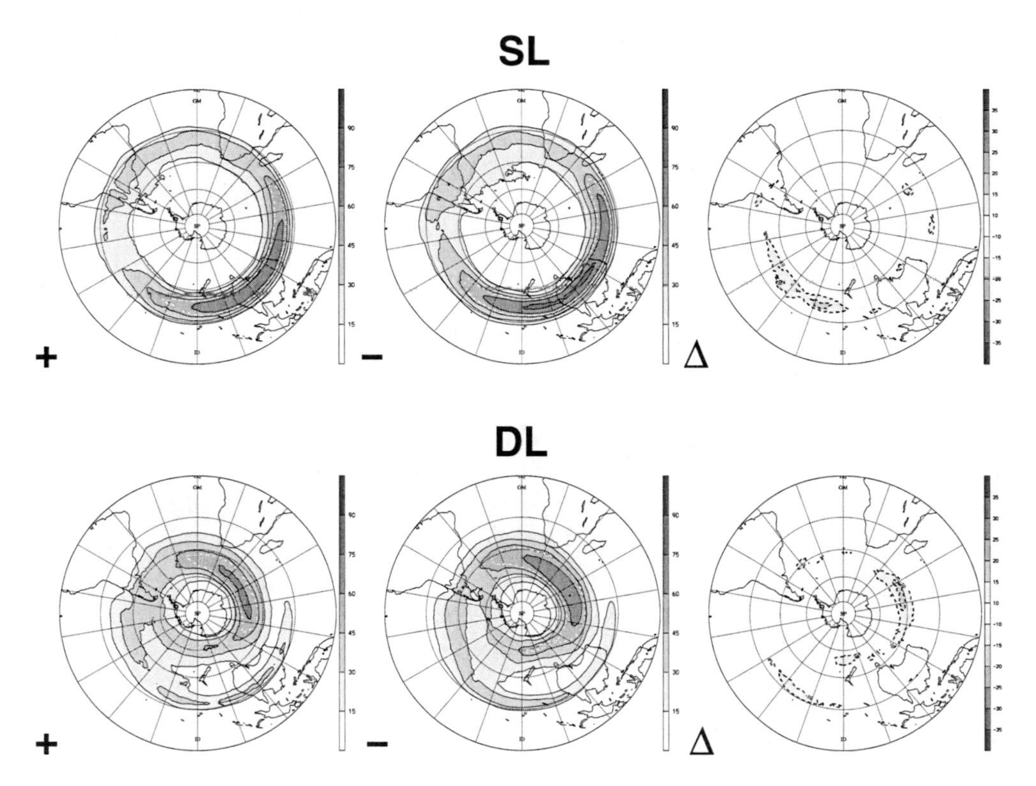

Figure 5.12: Same as Fig. 5.9, but for the SO.

of the difference patterns for the NAO and AO shows similar features over the Atlantic for both SL and DL categories. However, discrepancies in the DL and SL categories appear over the Pacific and North America when the negative PNA and positive AO phases (or vice-versa) are compared. The jet event frequency differences suggest a shift to the north over the Pacific of the jet events of the SL category, as well as larger frequencies over the southern United States in the positive phase of the PNA. For the DL category, a decrease of the frequencies in the Bering Sea is accompanied by a large increase of the frequencies in a band extending north of 20°N from the western Pacific to Hawaii in the positive PNA phase. Conversely, the AO displays a slight extension of the SL pattern north of Hawaii and larger frequency values from the Gulf of Mexico to the mid-Atlantic in the negative phase. For the DL category, although there is a decrease of the frequencies in the Bering Sea similar to the positive phase of the PNA, there is no signal in the frequency difference comparable to the large one found over the Pacific for the PNA.
This analysis shows that, although the AO and NAO share similar features over the Atlantic sector, the AO can not be simply considered as a combination of the NAO and PNA, as suggested in Section 4.3, where the variations in the SL

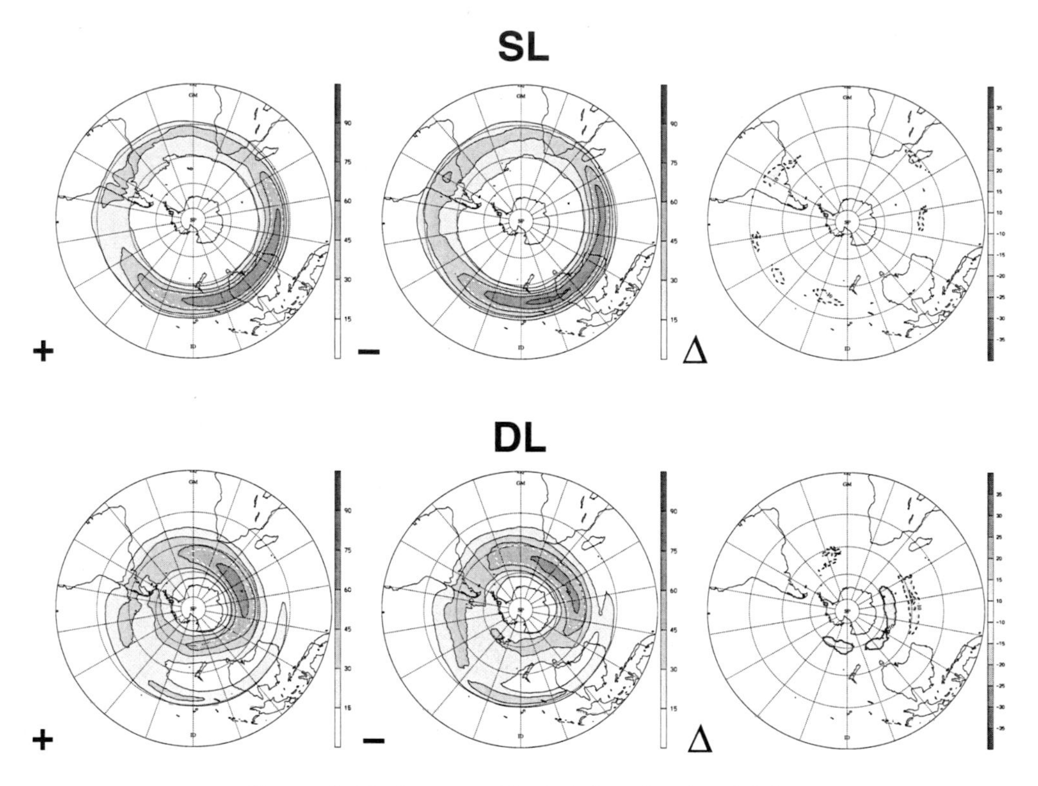

Figure 5.13: Same as Fig. 5.9, but for the AAO.

and DL categories are investigated. However, these elements show differences between the NAO, PNA and AO that raise further questions about the physical interpretations to give to the AO as an annular modulus (e.g. Ambaum et al. 2001). In effect, no signal in the jet event frequency difference, as clear as the one over the Atlantic for the DL category, has been found in the Pacific. In the case of an annular mode, signals of similar magnitude might be expected over a large part of the hemisphere and not only in very localized regions (in this case, the Atlantic sector). A relevant difference between the present study and previous studies is that the variations of salient synoptic features (jet streams) have been studied without the use of principal component analysis and EOFs. This gives a direct and more intuitive picture of the variations of upper-level jet patterns with the flow variability modes. These results also show that further studies are needed in order to better understand and assess annular moduli for the global atmospheric circulation, especially in the Northern Hemisphere.

5.6 Summary

In this Chapter, a study has been made of the vertical characteristics of the jet events determined in Chapter 4 by calculating the ratio of the decrease of their horizontal velocity in the 200-500 hPa layer to the wind speed at 200 hPa. A partitioning of the jet events has been made based on the evaluation of this ratio relative to a given threshold. On the one hand, ratios larger than the threshold are characterized by a strong vertical wind shear in the upper troposphere and an associated baroclinicity that is confined to upper levels. On the other hand, ratios smaller than the threshold denote a small vertical wind shear in the upper troposphere and an associated baroclinicity that extends farther down over a deep vertical layer. These two distinct groups of jet events have been designated as the shallow layer (SL) and deep layer (DL) baroclinicity categories.
Seasonal climatologies have revealed that jet events of the SL category are primarily found equatorward of the 40° latitude in both hemispheres with a slight poleward shift in summer and a large annual variability in amplitude. Jet events of the DL category are mainly found poleward of the 35° latitude with some occurrences in winter regrouped into a narrow band spiraling over half of the hemisphere in the 20-35° zonal belt. Note that the strongest frequencies have been found over oceans. The austral DL jet event pattern has a much weaker annual variability than its boreal counterpart. An analysis of the interannual variations of the jet event patterns of both categories in winter for the major modes of variability of the global circulation has also been made, permitting a refinement of the analysis made in section 4.3.
The SL and DL categories have been proposed as a novel way to describe the two types of upper-tropospheric jet stream designated in the literature by the terms of 'subtropical' and 'polar-front' jets. Also, a novel version of the schematic of Riehl (1962), depicting the mean winter positions of both the subtropical and polar-front jet streams, has been proposed for both hemispheres in this perspective. Finally, the correspondence between the positions of regions of high surface cyclone frequencies (i.e. storm tracks) and the DL jet events has been explored.

Chapter 6

Characteristics of Single and Double Jets

As previously mentioned in Chapter 3 and in Section 5.5.3, jet configurations can occur where two (or more) jets are found at a given longitude. In this Chapter a study is first presented of the zonal and geographical winter distributions of single- and double-jet events in both hemispheres. Second, a search is made for spatially coherent single- and double-jet events in particular regions and, finally, composites for each jet configuration are calculated. Various properties of the composites are investigated, including the baroclinic characteristics in the SL/DL perspective.

6.1 Identification

The global jet event climatology of Chapter 4 serves as the starting point for the retrieval of the meridional number of jets in both hemispheres. As the jet structure reaches its most developed state during winter, only the boreal and austral winters will be considered here.
The determination of the meridional jet configuration (single, double jets or more) is realized in both hemispheres by counting the number of distinct meridional segments formed by jet events along each longitude from the Equator up to a latitude of 80° (see Fig. 6.1). A segment is defined here as a meridional series of successive jet events at a given longitude. In order to be taken into account, a segment must have a minimum meridional extension of 4°. In the schematic representation of Fig. 6.1, the segment just north of the Equator is not taken into account as its meridional extension is less than 4°, while two other segments are found to the north. Thus the whole ensemble of winter jet events is decomposed

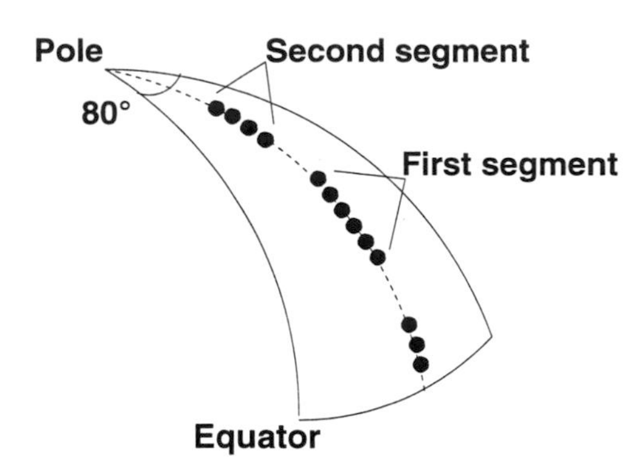

Figure 6.1: Schematic representation of the method used to count the jet segments on a given longitude (dotted line). The picture shows a portion of the Northern Hemisphere. Jet events are marked by black dots. The distance between two events in a segment is 1° latitude.

in the following way:

$$\mathcal{J}_G = \mathcal{J}_1 + \mathcal{J}_2 + \mathcal{J}_R \tag{6.1}$$

where $\mathcal{J}_G$ refers to the whole ensemble, $\mathcal{J}_1$ and $\mathcal{J}_2$ respectively to the jet events of the single-jet and double-jet categories. $\mathcal{J}_R$ regroups all the jet events belonging either to a jet configuration with a higher number of jets (triple and higher) or to jet events associated with segments meridionally too short to be accepted by the algorithm. Preliminary studies for triple-jet event configurations (similar to those proposed by Shapiro et al. 1987, e.g. Fig. 3.3 in Section 3.4) have shown very low frequencies (an order of magnitude less than those for the double-jets). Therefore this study will focus only on the single- and double-jet categories.

6.2 Winter Climatology

In this Section a search is made for the regions where single- or double-jets are found. The results of Chapters 4 and 5 suggest that single- and double-jets may occur in very specific regions and that differences between both hemispheres can be expected.

6.2.1 Zonal Distribution

A first estimate of the regions where single- or double-jets configurations dominate can be gained by calculating the zonal distribution of the occurrences of each category. Individual regions of enhanced frequencies appear clearly for each category, indicative of the inhomogeneity of the distribution of single and double jets around the globe. In the Northern Hemisphere (Fig. 6.2 a), two peaks in the distribution of the single jets are found at around 140°E (with a value of nearly 90%, meaning that in 90% of the winter time instances of the ERA15 period, jets in this region have a meridional single-jet configuration), a second one

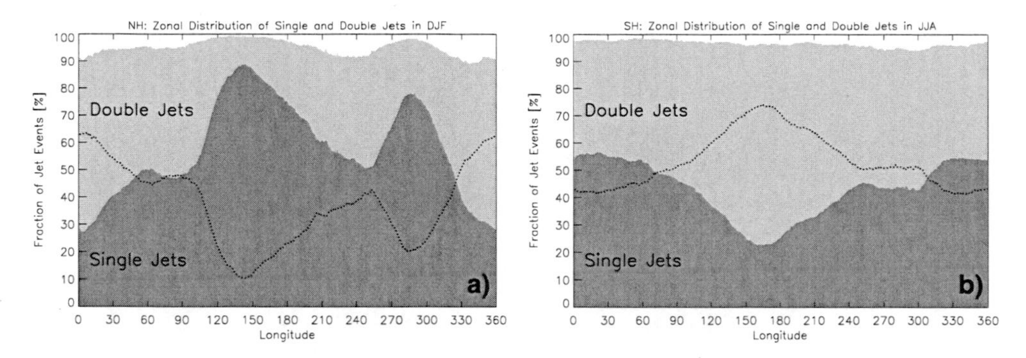

Figure 6.2: Mean winter zonal distribution in % of all jet events ($\mathcal{J}_G$ in Equation 6.1) partitioned between the single jet category ($\mathcal{J}_1$, dark grey region), the double jet category ($\mathcal{J}_2$, light grey region) and the rest ($\mathcal{J}_R$, white region) for (a) the Northern Hemisphere and (b) the Southern Hemisphere. The dashed black line corresponds to the double-jet event fraction alone, as discussed in Section 6.3.1

around 290°E (with a value reaching 80%) and a much smaller third one (50%) at around 60°E. Except for a region comprised roughly between 40°W and 100°E, the fraction of winter single jet events is larger than 50%, suggesting on average a predominance of the meridional single-jet state for the Northern Hemisphere. Double jet events appear to be the most commonly found from about 30°W and 40°E, with a maximum near 5°E. The fraction of jet events that do not belong to the single or double jet categories does not exceed 10% of the total amount of events.
The situation is quite different in the Southern Hemisphere (Fig. 6.2 b). There is no large localised maximum in the distribution of single-jet events but a clear minimum (of less than 25%) located near 170°E. The fraction of single-jet events is slightly larger than 50% from 40°W to 90°E and double jets clearly dominate outside this interval.

6.2.2 Geographical Distribution

Figure 6.3 shows the geographical distributions of single and double jets.
Single jets: The Northern Hemisphere (Fig. 6.3 a) exhibits a spiral-like structure (analogous to the one found for the total winter jet event climatology in Fig. 4.2 a) starting over central North Africa and ending over the eastern Atlantic, mainly in the 20°N-50°N belt. Maximum frequencies are found over the western Pacific and eastern North American sectors. Note also lower frequencies in the eastern Pacific and western North American sectors, as well as in the eastern Atlantic-western European sector. The situation is quite different in the Southern Hemisphere. There, the single-jet pattern exhibits a quasi-axisymmetric

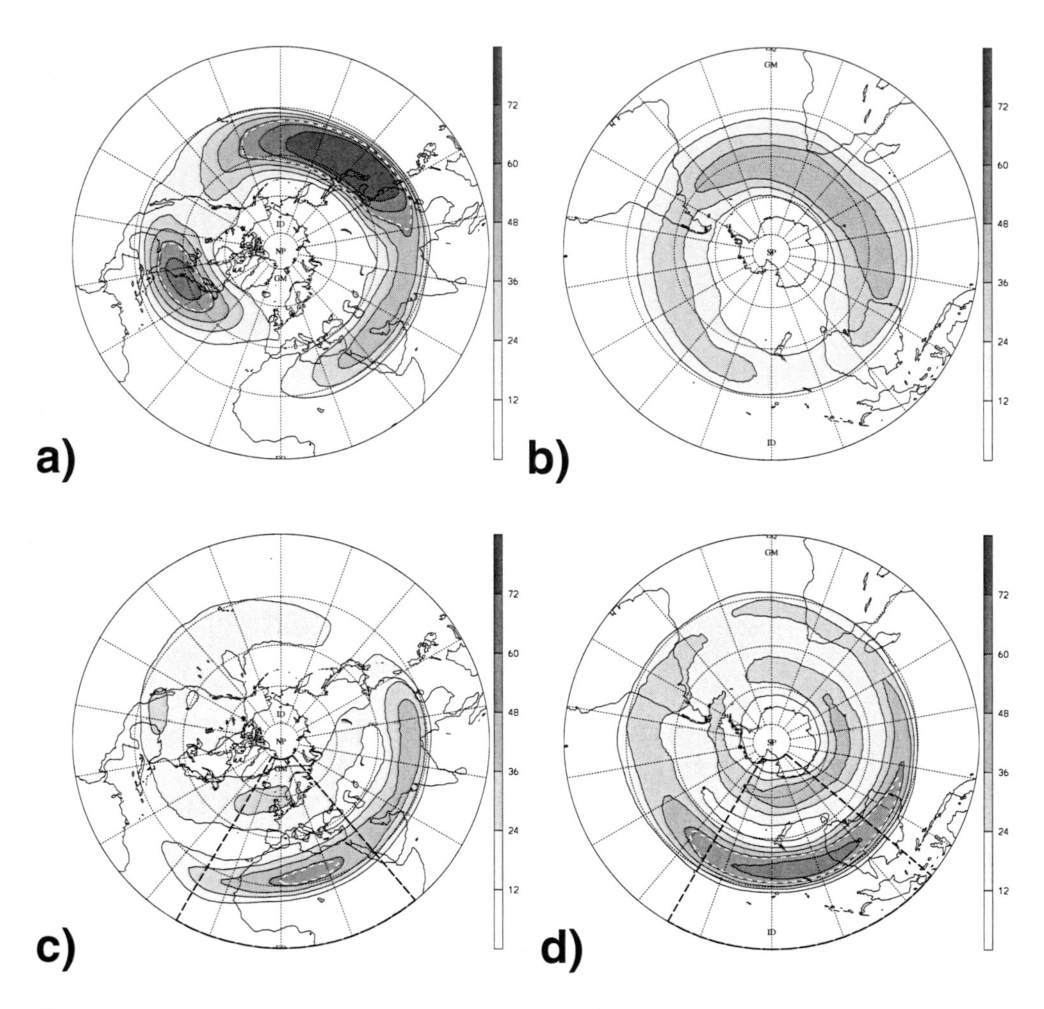

Figure 6.3: Mean winter positions of single jets (a and b) and double jets (c and d) in the Northern Hemisphere (left panels) and the Southern Hemisphere (right panels). Frequencies in %, the white dotted line denoting the 50% frequency. Note the slightly different scale compared to the similar figures presented in the former two chapters. The longdashed rectangles in (c) and (d) delimits the domains discussed in Section 6.3.1.

ring-shape with maximum values from the Atlantic to the southwest of Australia (see Fig. 6.3 b). The meridional width of the ring pattern is not constant as jet events occur from 25°S to 60°S from the Atlantic to the mid-Indian Ocean and mainly in the 20°S-40°S in the vicinity of New Zealand. As already stated earlier, this latter region is also the area where single-jet events are the most rare.

Double jets: The pattern of double-jet events in the Northern Hemisphere is composed of two features (Fig. 6.3 c): a long zonally extended pattern from the mid-Atlantic south of 20°N to the China Sea and a meridionally irregular pattern that extends from the mid-Pacific to the north of Mongolia. The Southern Hemisphere is characterized by frequencies exceeding 12% poleward of 20°S and a narrow void region from Tasmania to the mid-Pacific (Fig. 6.3 d). Inside this large feature, two concentric ring patterns are distinguishable. The equatorward ring is inside the 20°S-40°S belt from the mid-Atlantic to the eastern coasts of South America. The poleward ring pattern is located in the 45°S-70°S belt and extends from the mid-Atlantic to the northeast of the Antarctic Peninsula.

6.2.3 Discussion

In this Section, the winter distributions of single- and double-jet events have been presented. Single jets dominate the palette of winter upper-level flow configurations over the major part of the Northern Hemisphere. Regions with particularly high rates of single-jet occurrences are located in the midlatitudes in the eastern Asian-western Pacific sector and the eastern North American-western Atlantic area. Following the results of Chapter 5, these regions are characterized by the superposition of shallow and deep-layer baroclinic jet event patterns (e.g. Fig. 5.6 a). Double-jet events exist in the Northern Hemisphere and are found in particular in the eastern Atlantic-Eurasian sector. In this sector, two well separated patterns of jet event are found: one centered near 25°N and the second between 45°N and 50°N. Frequent double-jet events are also found from the eastern Pacific to the Rockies. However, in this region, the identification of two meridionally separated patterns is not obvious.
In the Southern Hemisphere, the repartition of the single- and double-jet configurations is balanced between the mid-Pacific and the southeastern Indian Ocean with frequencies for both types of flows ranging from 45% to 55%. However, the region comprised Australia and the mid-Pacific is dominated by the double-jet flow configuration. Note that the split flow studied by Bals-Elsholz et al. (2001) is precisely located on the western part of this region. In contrast to the Northern Hemisphere, double-jet event patterns in the Southern Hemisphere exhibit two meridionally well separated concentric rings over most of the hemisphere.
Studies of the atmospheric circulations and jet configurations of other planets can be quite instructive in order to better understand the characteristics of the global atmospheric circulation of the Earth. Williams (1978, 1979a, 1979b, 1988) compared different planetary circulations for the Jovian and terrestrial atmospheres. He demonstrated among other things the relevance of the size of the planet, as well as of the rotational rate and surface drag on setting the multiple-jet configurations. Thus for the Jovian atmosphere, multiple, highly zonal jets prevail due to the great size and high rotational rate of Jupiter. The existing terrestrial circulation is in a transient state close to a double-jet system. Using

a quasi-geostrophic model Williams (1979b) managed to create regular multiple jets by increasing the Earth rotational rate (four times the actual Earth rotation) and by reducing the surface drag. The results of the present study show the distinct mean winter geographical distribution of the double-jet events in the Southern Hemisphere where the fraction of land mass is low. On the contrary, the results for the Northern Hemisphere suggest that forcing from land masses and from topography can alter the flow characteristics and lead to other flow configurations (i.e. the predominance of the single-jet configuration over most of the hemisphere).
Note finally that no conclusion can be drawn from these results concerning the zonal length and time duration of single and double jets. Indeed, the search for multiple-jet occurrences has been made at each time instance and on each longitude *separately*, without taking into account the nearby time instances and longitudes. Elements related to the duration and the zonal coherency of single and double jets will be discussed in the following Section.

6.3 Composites

As stated earlier, the results of Section 6.2 bear no indication on the temporal and spatial coherency of single and double jets. In this Section, a search is made for zonally coherent double-jet events. The regions where double jets frequently occur (already identified in Section 6.2.1) are chosen in both hemispheres for this investigation. Single-jet events are also searched in these regions. The aim is to build composites of the mean properties of the troposphere and lower stratosphere when single- or double-jet events occur. These two mean states can then serve as an indicator of two 'extreme' states between which most of the winter time instances are likely to be found. Special attention will be given to baroclinic properties and to further features such as storm tracks and mean sea-level pressure.

6.3.1 Methodology

Double-jet and single-jet events are searched from the results of Section 6.2 for each winter time instance in particular regions. The determination of these regions is made by considering zonal intervals in Figs. 6.2 (a) and (b) where double jets dominate, e.g. where the double-jet frequency exceeds 50%. This corresponds to the mid-Atlantic-European sector in the Northern Hemisphere and to a good half of the Southern Hemisphere. A further refinement of both intervals is made by choosing similar zonal extensions in order to permit further comparisons between both hemispheres. The intervals are therefore chosen from 30°W to 40°E and from 130°E to 210°E (region corresponding to double-jet frequencies larger than 65% in Fig. 6.2 b) for the Northern and Southern Hemispheres, respectively (see Figs. 6.3 c and d). Moreover, the interval must not be too long otherwise the condition upon the zonal extension of the multiple-jet event might prove to be too restrictive. Thus, the single or double-jet structure has to span continuously over more than 60% of the interval length in order to be considered. This corresponds to zonal distances at a latitude of 45° of nearly 3300 km and 3770 km in the Northern and in the Southern Hemisphere, respectively. Each winter time instance with a single- or double-jet event fulfilling this criterion is recorded. The total number of time instances obtained with this method is detailed month by month in the Tables C.1 and C.2.

6.3.2 Tropopause Characteristics

Northern Hemisphere

Differences between the single-jet and double-jet composites arise at the level of the dynamic tropopause (DT hereafter, defined as the 2-pvu surface). For the single-jet composite, one zonally elongated band of wind speeds higher than 30 ms^{-1} is found from the western Atlantic to the eastern Mediterranean (Fig. 6.4 a) with values exceeding 40 ms^{-1} west of Spain. The flow is mainly zonal over the Atlantic. On the contrary, two high wind speed regions are found in the double-jet composite (Fig. 6.4 b): the first one from the western Atlantic to Scandinavia, and the second one from west of Mauritania to the Red Sea. Low wind speeds (less than 20 ms^{-1}) are found on average west of and over Spain. The flow exhibits a split in the mid-Atlantic, with one branch turning northeastward and the second one decelerating southeastward, before turning eastward and accelerating south of 25°N. It is interesting to note that in this configuration, the flow at the DT is on average northwesterly in the midlatitudes from the mid-Atlantic to eastern Europe.
Similar differences are found for the potential temperature (θ) at the DT level (Figs. 6.4 c and d). The single-jet composite exhibits a zonal and intense meridional θ-gradient over the Atlantic and Europe. In the double-jet composite, two large meridional θ-gradients are present; one extends from Quebec to Scandinavia and the other one from the southeastern United States to the subtropical mid-Atlantic. Between these two regions, the gradient of potential temperature becomes very weak. Moreover, the potential temperature is higher over western Europe than in the single-jet composite, suggesting on average a higher tropopause in this region for the double-jet configuration.
This latter fact is confirmed by the inspection of the mean pressure on the DT. The double-jet composite (Fig. 6.4 f) is characterized by a relatively high DT

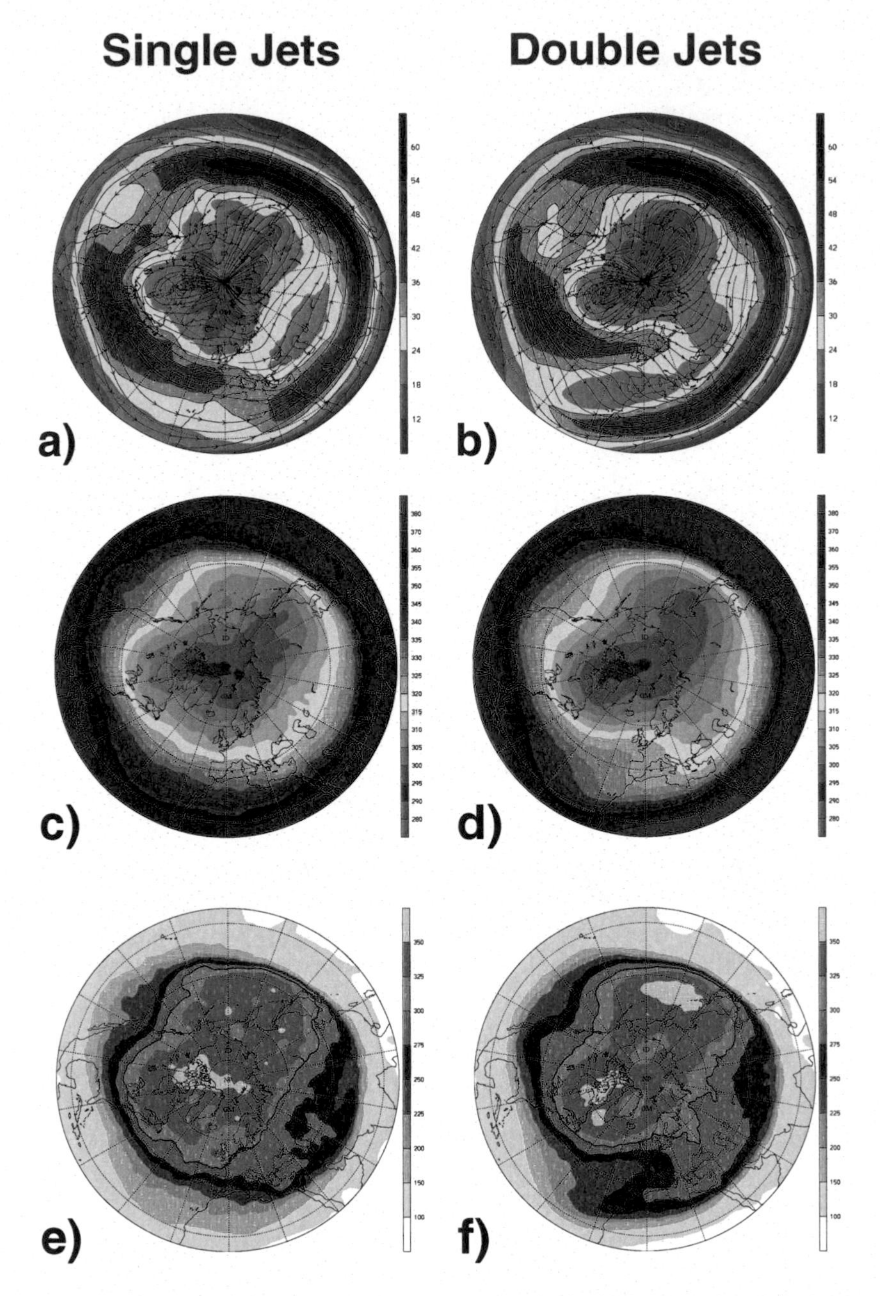

Figure 6.4: Composites of single-jet (left panels) and double-jet events (right panels) spanning over at least 60% of the interval from 30°W to 40°E in the Northern Hemisphere. Panels (a) and (b) show the horizontal wind velocity (in ms^{-1}) and streamlines (black lines) on the dynamic tropopause (defined as the 2-pvu surface), (c) and (d) the potential temperature in K on the dynamic tropopause (from 280 K to 380 K, contour interval 5 K between 290 K and 360 K) and (e) and (f) the mean pressure on the dynamic tropopause in hPa with the bold black line outlining 300 hPa.

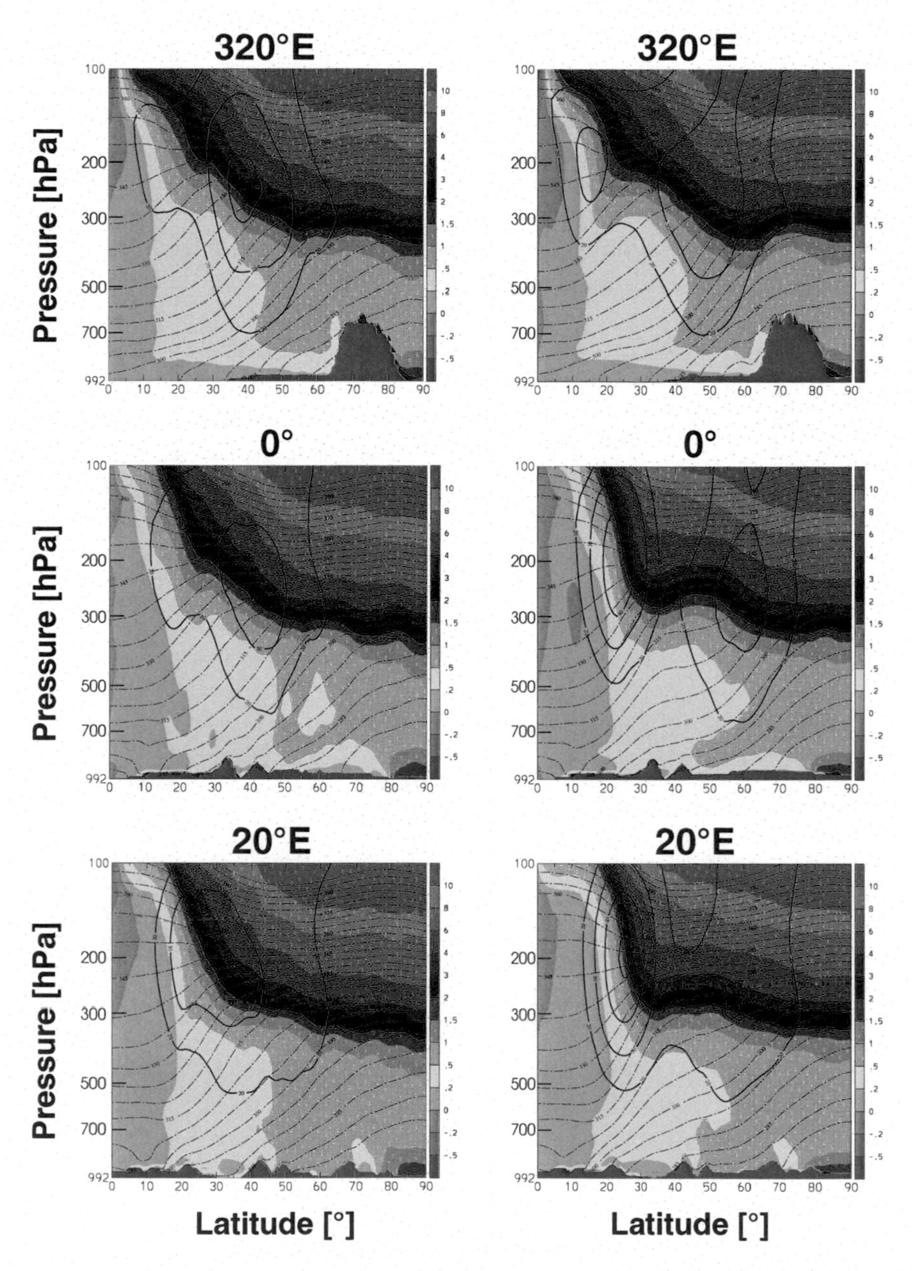

Figure 6.5: Cross sections for the Northern Hemisphere at 320°E, 0° and 20°E of PV (colors, in pvu, 2 pvu outlined by a black solid line), potential temperature (dashdotted, contour interval 5 K) and horizontal wind speed (bold, contour interval 10 ms^{-1} from 20 ms^{-1}) for the single-jet (left panels) and double-jet composites (right panels).

(low pressures on the DT) over the eastern Atlantic-European sector and a relatively low DT (high pressures on the DT) over northern Algeria. The large mean zonal pressure gradient around 40°N over eastern North America and the western Atlantic appears to split over the mid-Atlantic into two branches: one oriented southwest-to-northeast extends to Scandinavia and the other one oriented northwest-to-southeast extends to 20°N with lower wind speeds over the mid-Atlantic. Conversely, a continuous zonal pressure gradient extends around 40°N from eastern North America to Europe in the single-jet composite (Fig. 6.4 e). Consequently, the DT appears to be relatively low over the eastern Atlantic and western Europe sectors and high over the southern portion of the Mediterranean. No enhanced pressure gradient is found on the DT over the Sahara.
Interesting characteristics of the tropopause in both the single- and double-jet configurations can be obtained from vertical cross sections (see Fig. 6.5). At 320°E, two distinct jet cores with wind velocities exceeding 30 ms^{-1} are found in the double-jet configuration near 15°N and 50°N, respectively. The southernmost jet core is located at nearly 200 hPa while the northernmost core is extending from 500 hPa upward into the stratosphere. In comparison, the single-jet configuration exhibits one jet core with wind speeds exceeding 40 ms^{-1} near 40°N and between 200 and 300 hPa. Except for a slightly lower tropopause near 60°N in the double-jet composite, no relevant difference in the tropopause height is visible between both configurations.
Two distinct jet cores are found at 0° in the double-jet configuration. The southernmost jet core is centered near 25°N and at 200 hPa and the wind speed is over 40 ms^{-1} while the northern most jet core is centered a little lower (at about 250 hPa) near 60°N with lower wind speeds (not exceeding 40 ms^{-1}). The tropopause is very steep in the subtropics and drops to nearly 300 hPa at 35°N. A second drop in the height of the tropopause is found poleward of 55°N. Between these two drops, the tropopause is higher and exibits a dome-like structure. In the single-jet configuration, the jet core is centered at nearly 40°N and at the same level as the northern most jet core in the double-jet composite. The wind speed is lower (not exceeding 40 ms^{-1}). The tropopause is characterized by a single continuous drop and is less steep in the subtropics.
At 20°E, the northernmost jet core has shrunk and is located at nearly 55°N at a height of about 250 hPa in the double-jet configuration. The southernmost jet core is still located at 25°N and at a height of 200 hPa but has grown stronger in comparison with the 0° cross section (wind speeds over 50 ms^{-1}). A steep tropopause is still visible in the subtropics and the dome-like structure of the tropopause in the midlatitudes has nearly vanished. Instead, a smooth and continuous drop occurs poleward of 50°N. The jet core of the single-jet configuration has moved southward to nearly 30°N at a height of 200 hPa. The wind speed does not exceed 40 ms^{-1}. The tropopause is less steep in the subtropics than in similar regions in the double-jet composite.

Southern Hemisphere

In the single-jet composite, wind speeds at the DT level (defined as the -2-pvu surface) are low in the midlatitudes south of Australia and New Zealand and are high east of Australia (Fig. 6.6 a). In comparison, the double-jet composite exhibits large wind velocity south of Australia and New Zealand and slightly weaker ones east of Australia (see Fig. 6.6 b). These two patterns of wind speeds larger than 30 ms^{-1} are separated by a zone of lower wind speeds (not exceeding 20 ms^{-1}) extending from Tasmania to New Zealand. The mean flow is also quite different. In the single-jet configuration, the streamlines are mainly southwesterly oriented south of Australia, so that confluence is present on average over southeastern Australia. In the double-jet composite, there is a split of the flow in the Indian Ocean southwest of Australia with one branch being equatorward of 35°S and the second poleward of nearly 50°S.
The potential temperature at the DT is clearly lower over the Tasman Sea in the single-jet configuration (differences up to 10 K) compared to the double-jet composite. In the single-jet configuration, a trough of the 310 K isentrope is visible on Fig. 6.6 (c). Conversely, the 310 K isentrope of the double-jet composite (Fig. 6.6 d) has a wavelike structure, with a weak ridge and a nearly zonal pattern over the Tasman Sea and a pronounced trough southeast of New Zealand (the trough axis being near 200°E). Due to this difference, the θ-gradient is more pronounced south of Australia and New Zealand in the double-jet configuration than in the single-jet configuration.
The characteristics of the potential temperature on the DT described above find their equivalent features in the mean pressure distribution on the DT. The single-jet composite (Fig. 6.6 e) is characterized by a quasi-uniform height over the western South Pacific poleward of New Zealand. However the DT appears to be locally lower from Tasmania to east of New Zealand. This in turn enhances the pressure gradient on the DT around 30°S from eastern Australia to the western Pacific east of New Zealand. Conversely in the double-jet composite (Fig. 6.6 f) the DT is relatively low poleward of 60°S and relatively high in an isolated region southwest of New Zealand. This particular topography of the DT implies an enhancement of the pressure gradient around 60°S. It is also interesting to notice that the DT is higher around 30°S from Australia to the western Pacific than in the single-jet composite, diminishing the pressure gradient in this region.
Both cross sections at 120°E (Fig. 6.7) exhibit a jet core that is located at nearly the same latitude (between 25°S and 30°S), at the same height (slightly beneath 200 hPa) and with the same strength (wind speed exceeding 50 ms^{-1}). A second jet core is visible in both configurations in the midlatitudes. However, in the double-jet composite, the jet core of 30 ms^{-1} is extending from nearly 400 hPa upward into the stratosphere while the single-jet composite exhibits only a vertically confined core near 300 hPa. The tropopause characteristics are also similar

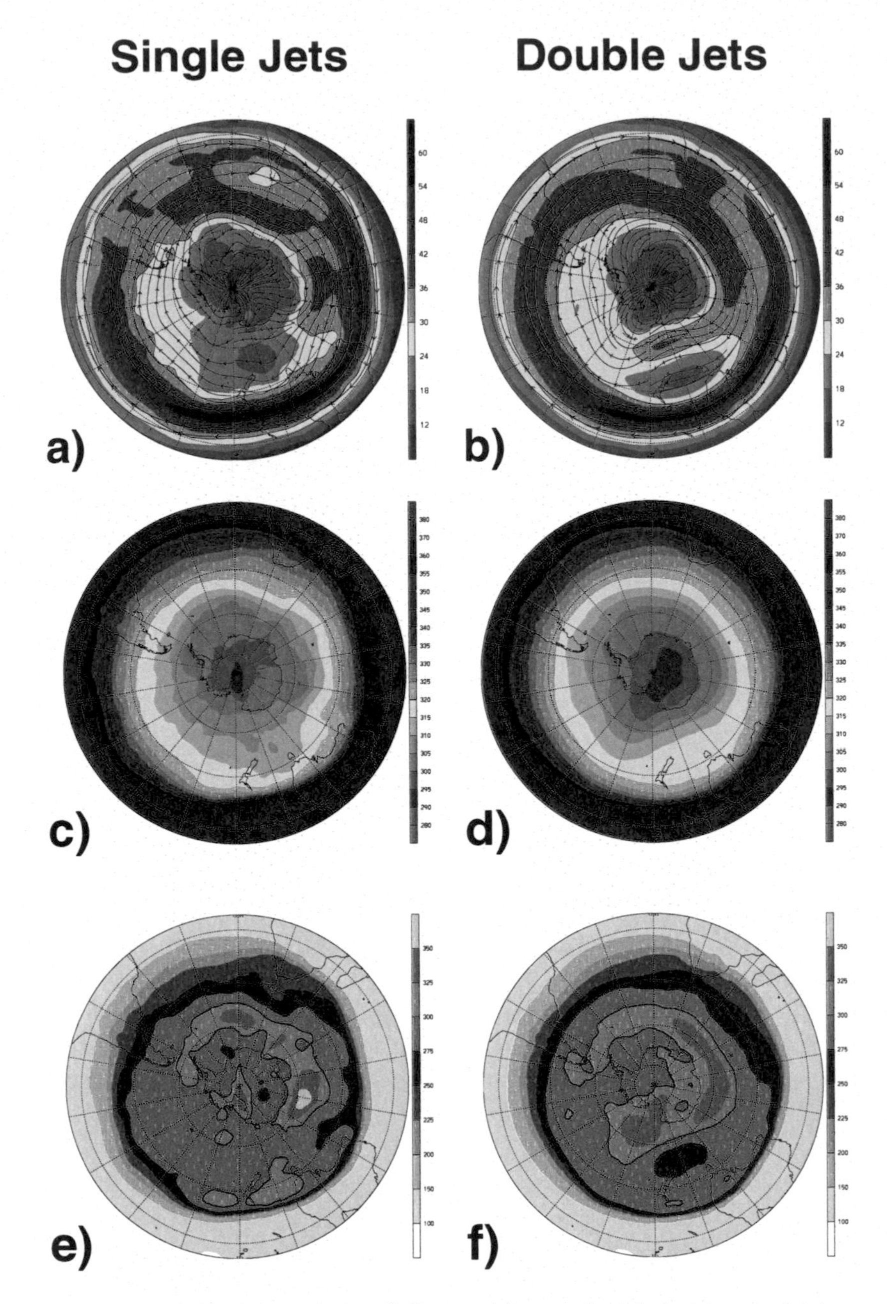

Figure 6.6: Composites of single-jet (left panels) and double-jet events (right panels) spanning over at least 60% of the interval from 130°E to 210°E in the Southern Hemisphere. Panels (a) and (b) show the horizontal wind velocity (in ms^{-1}) and streamlines (black lines) on the dynamic tropopause, (c) and (d) the potential temperature in K on the dynamic tropopause (from 280 K to 380 K, contour interval 5 K between 290 K and 360 K) and (e) and (f) the mean pressure on the dynamic tropopause in hPa with the bold black line outlining 300 hPa.

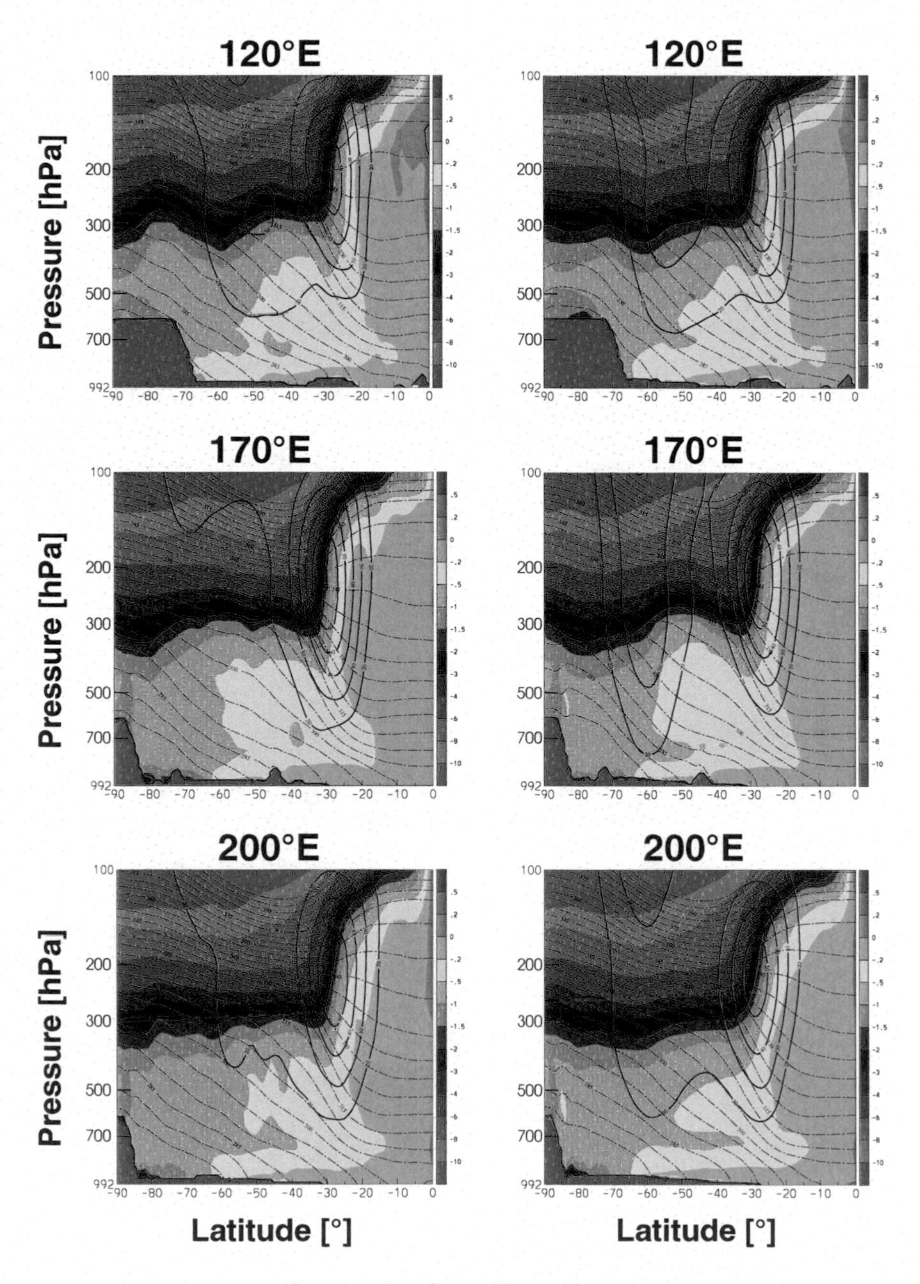

Figure 6.7: Cross sections for the Southern Hemisphere at 120°E, 170°E and 200°E of PV (colors, in pvu, -2 pvu outlined by a black solid line), potential temperature (dashdotted, contour interval 5 K) and horizontal wind speed (bold, contour interval 10 ms^{-1} from 20 ms^{-1}) for the single-jet (left panels) and double-jet composites (right panels).

with a steep drop of height in the subtropics. A slightly higher tropopause is visible in the single-jet composite near 75°S.
At 170°E two jet cores are depicted in the double-jet configurations. The northernmost jet is located near 30°S and at a height of about 200 hPa. The wind speed exceeds 50 ms^{-1}. The southernmost jet core extends vertically from 500 hPa upward into the stratosphere with wind speeds exceeding 30 ms^{-1}. In the single-jet configuration, the jet located around 60°S has disappeared and the jet core to the south is present at a latitude and height similar to the double-jet composite with a larger wind speed (exceeding 70 ms^{-1}). The tropopause behaviour in both configurations is similar in the subtropics. In the midlatitudes, the double-jet composite is characterized by a slightly higher tropopause and a second drop at nearly 60°S.
The cross sections at 200°E are similar to those at 120°E, except that no distinct jet core is found at the tropopause level poleward of the northernmost jet. Note that this jet is similar in position and strength to the one at 120°E in position and strength. The tropopause characteristics are similar in both configurations with a steep tropopause in the subtropics and a smooth tropopause poleward of 30°S.

Interhemispheric Comparison

Some fundamental differences between both hemispheres regarding the synoptic structures accompanying single and double jets arise from the analysis of the tropopause characteristics.
The flow at tropopause level in the single-jet configuration in the Northern Hemisphere over the Atlantic is on average zonal. In the Southern Hemisphere, the single-jet composite is characterized by a southwesterly flow in the eastern Indian Ocean from the midlatitudes into the subtropics. There is a confluence of this flow with the subtropical zonal flow over southeastern Australia. In the double-jet configuration in the Northern Hemisphere, there is a split of the jet in the mid-Atlantic, with a first branch turning northeastward and a second one turning southeastward. The flow downstream of the split is on average northwesterly. In the Southern Hemisphere, the double-jet configuration is characterized by a split of the flow over the eastern Indian Ocean with the poleward branch turning southeastward to 60°S. On average, both branches feature zonal flows, and no equatorward flow comparable to the double-jet composite in the Northern Hemisphere can be identified.
The vertical structure of the tropopause in the Northern Hemisphere is clearly zonal in the midlatitudes in the single-jet configuration as suggested by the zonal pressure gradient around 40°N. The Southern Hemisphere is characterized by a relatively flat tropopause over the western Pacific with an anomalously low height from Tasmania to east of New Zealand around 30°S. The Northern Hemi-

sphere exhibits a wave-like pattern in the double-jet configuration with a higher tropopause over the eastern Atlantic and a lower one over northern Algeria. In the Southern Hemisphere, the double-jet composite is characterized by an isolated region with a high tropopause south of Tasmania and New Zealand and a low tropopause poleward of 60°S.
The mean meridional position of the single jet in the Northern Hemisphere (near 40°N) is nearly intermediate compared to the meridional positions of the subtropical and polar-front jets of the double-jet configuration (25°N and poleward of 50°N, respectively). In the Southern Hemisphere, the disappearance of the southernmost jet does not modifiy the position of the northern most jet that remains steadily centered near 25°S (compare Figs. 6.4 and 6.6).
In the Northern Hemisphere, the single-jet composite for the Northern Hemisphere exhibits on average a jet 10 ms^{-1} weaker than the southernmost jet in the double-jet configuration. In the Southern Hemisphere, the wind speed of the jet around 25°S is on average 20 ms^{-1} larger in the single-jet than in the double-jet configuration. This result confirms and generalizes previous results obtained by Chen et al. (1996) in their study of the split jet during the 1986-1989 ENSO cycle.

6.3.3 Baroclinic and Surface Pattern Characteristics

Differences at tropopause level between the single- and double-jet composites of Section 6.3.2 are accompanied by major differences in the associated baroclinicity characteristics and mean positions of surface transient synoptic systems. In this Section, the distinction between the deep-layer (DL) and shallow-layer (SL) baroclinicity categories developed in Chapter 5 is applied to the single- and double-jet composites. Mean surface patterns are studied in the form of charts of mean sea-level pressure and mean frequencies of moving surface cyclones are calculated using the method developed by Wernli and Schwierz (2004) (e.g. Section 5.4) and compared for each type of jet configuration.

Baroclinic Characteristics

In the Northern Hemisphere, the composites of single-jet events occurring in the Northern Atlantic and European sectors are characterized by a zonal band pattern of DL jet events extending from the the western Atlantic to the Mediterranean with the maximum frequencies over the western Atlantic and Spain (Fig. 6.8 a). In the double-jet configuration, the DL jet event pattern has a southwest-to-northeast orientation from the Great Lakes to Scotland with maximum frequencies east of Nova Scotia (Fig. 6.8 b). A comparison with the results of

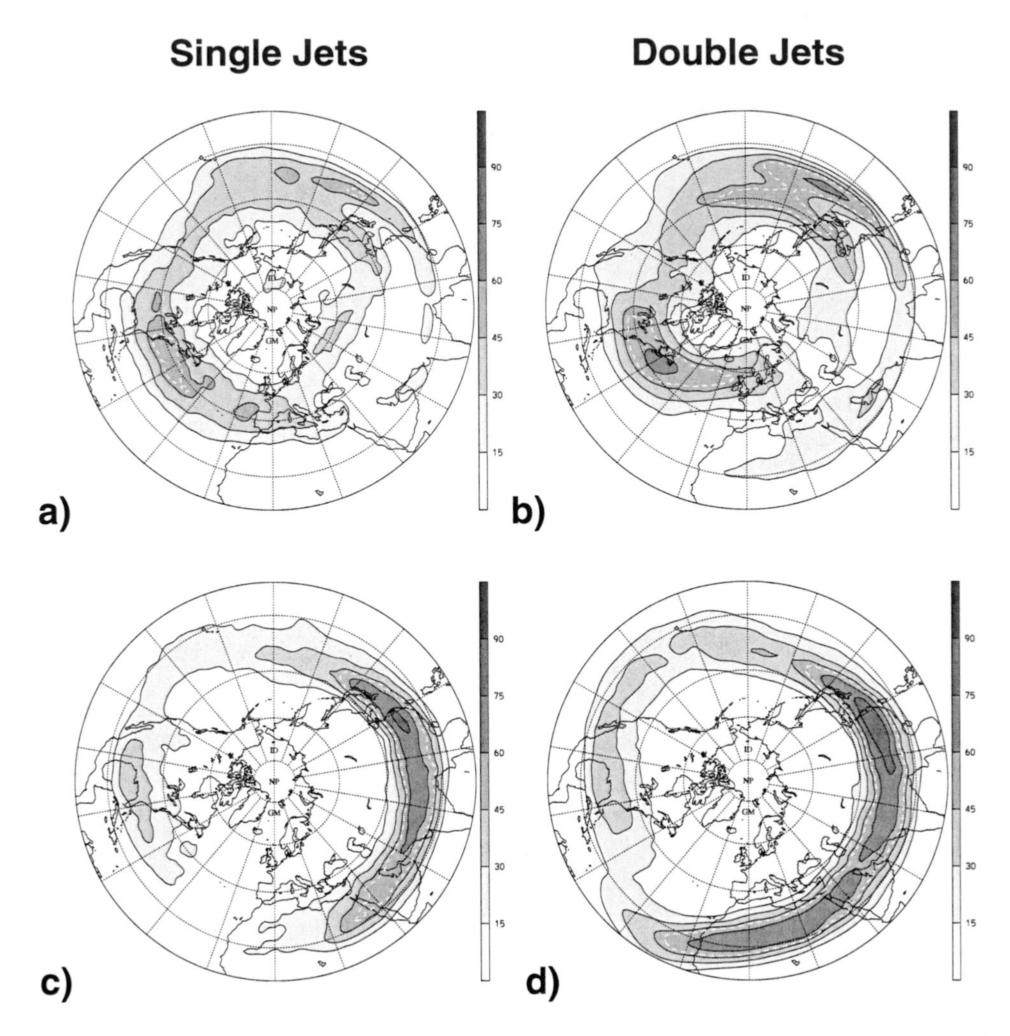

Figure 6.8: Jet event frequencies in % for the DL (a and b) and SL categories (c and d) for the single-jet (left panels) and double-jet (right panels) composites in the Northern Hemisphere, calculated with the same ratio used in Chapter 5 (0.4). The dashed white line outlines 50%.

Section 6.3.2 indicates clearly that the climatological single jet over the Atlantic centered around 40°N is mainly associated with a deep-layer baroclinicity. In the single-jet composite, the Atlantic and western part of Sahara are void of SL jet events (Fig. 6.8 c). Conversely, the double-jet composite displays SL jet events from the mid-Atlantic to the Red Sea (Fig. 6.8 d) and the whole pattern is comparable to the one found in the total winter climatology (e.g. Fig. 5.2 DJF).
In the Southern Hemisphere, the single-jet composite displays no DL jet frequency pattern around 60°S from the eastern Indian Ocean to the western Pacific

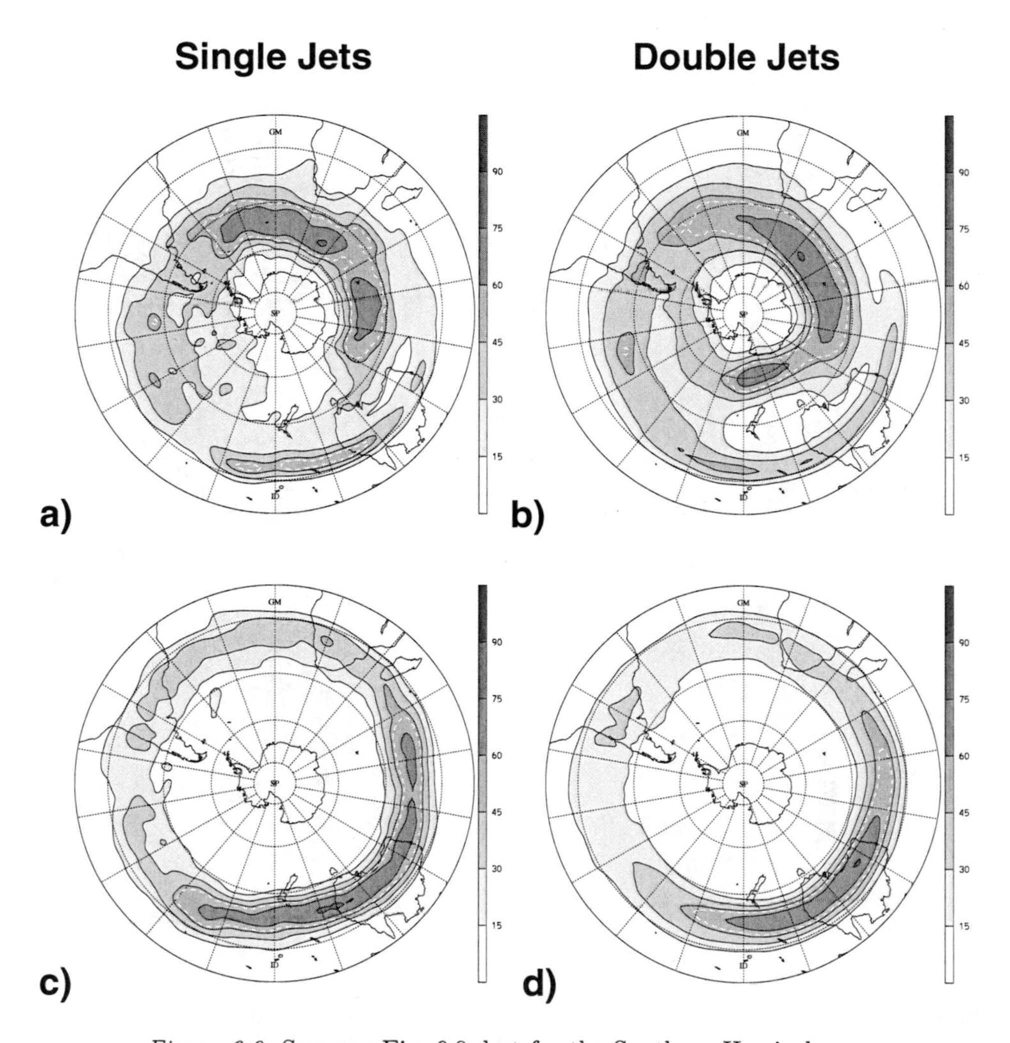

Figure 6.9: Same as Fig. 6.8, but for the Southern Hemisphere.

(Fig. 6.9 a). Conversely, the double-jet composite shows large frequencies of DL jet events poleward of 50°S from the Atlantic to the western Pacific (Fig. 6.9 b), giving to the whole DL jet event pattern a spiral-like shape comparable to the one found in the winter climatology (e.g. Fig. 5.5 c). The patterns of SL jet events are similar in both configurations, i.e. they exhibit a ring-like pattern in the 20°S-40°S belt (Figs. 6.9 c and d). Slight differences are nonetheless noticeable in the intensities within the SL jet frequency patterns. Slightly larger SL jet event frequencies extend from Australia to Madagascar and from north of New Zealand to the mid-Pacific in the single-jet configuration.
Major interhemispheric differences in the baroclinic characteristics of single and double jets arise. First, the SL jet event frequency pattern's variability is large

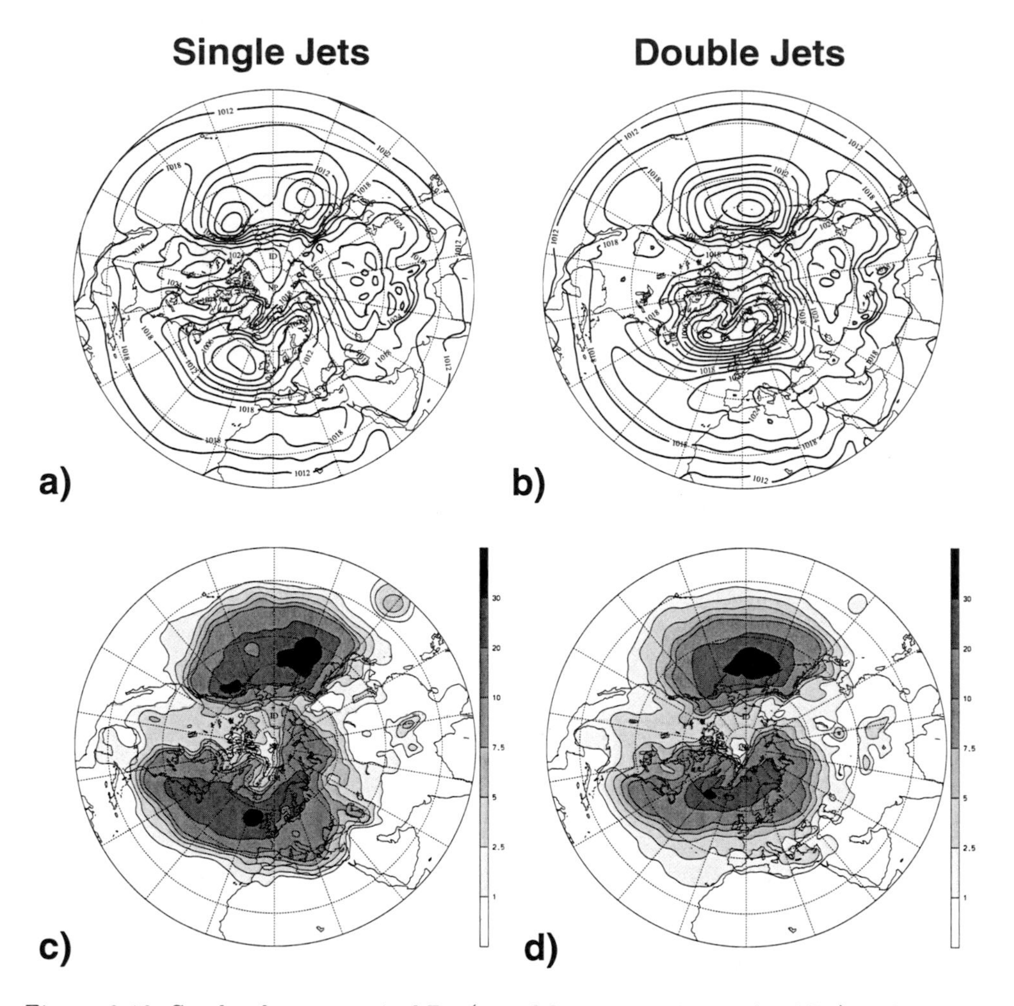

Figure 6.10: Sea-level pressure in hPa (a and b, contour interval 3 hPa) and moving surface cyclone frequencies in % (c and d) for the single-jets (left panels) and double-jets (right panels) composites in the Northern Hemisphere. Contours for panels (c) and (d) are the same as in Fig. 5.8 (c) and (d).

over the Atlantic in the Northern Hemisphere. In the Southern Hemisphere the variability, is quite small, except for slight differences in the intensities within the SL pattern. Second, in the Northern Hemisphere, the DL jet pattern is zonally (southwest-to-northeast) oriented in the single-jet (double-jet) composite. The situation is quite different in the Southern Hemisphere in the single-jet composite. The DL jet pattern adopts a southwest-to-northeast orientation south of Australia and another pattern is present around 25°S east of Australia. This fact could suggest a behavior of the DL jet events comparable to the one described by Bals-Elsholz et al. (2001) for the polar-front jet in their study of single-jet

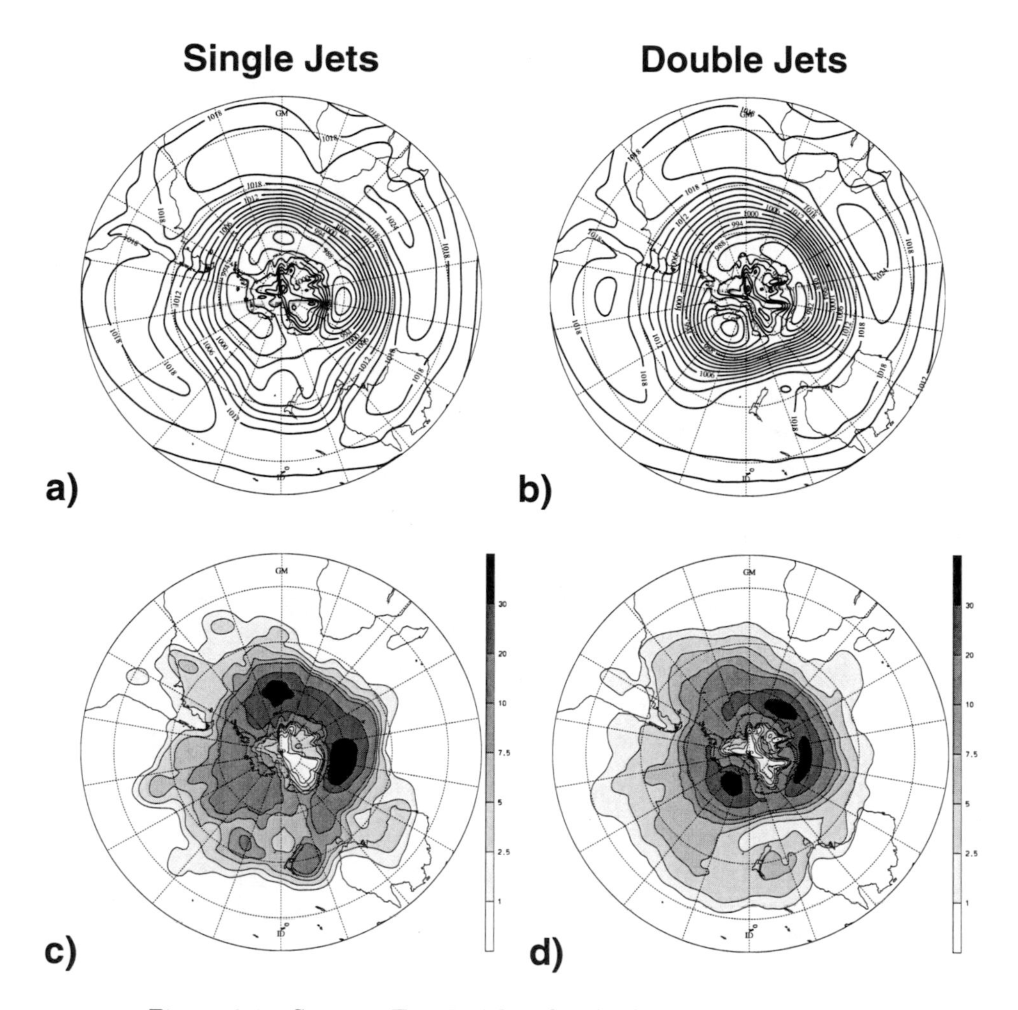

Figure 6.11: Same as Fig. 6.10 but for the Southern Hemisphere.

episodes in this precise region, i.e. an equatorward displacement of the jet due to a cold air surge from Antarctica south of New Zealand.

Surface Patterns

The time-mean sea-level pressure over the Atlantic in the Northern Hemisphere for the single-jet composite (Fig. 6.10 a) is characterized by a large low pressure system located west of Ireland and a zonally elongated high pressure system extending from Florida to Egypt. Between these two systems, a pressure gradient extends zonally around 40°N from the western Atlantic to western Europe. The double-jet composite (Fig. 6.10 b) shows a deep (994 hPa) low pressure system that extends from Greenland to the Barents Sea with two centers west of Ireland and north of Scandinavia, respectively and a strong (1024 hPa) high pressure system centered over western Europe and extending upstream over the Atlantic. Between these two features a strong pressure gradient extends from the Labrador Sea to the Baltic Sea. The two aforementioned pressure gradients are collocated with the DL jet event patterns previously described. Additionally, some differences are also visible in the Pacific. In the single-jet configuration, two lows are found in the time-mean sea-level pressure south of Kamtchatka and in the Gulf of Alaska, respectively with a sea-level pressure of 1003 hPa each. In the double-jet composite, a single deeper low (997 hPa) is found east of the Kouril Islands.

The situation in the Southern Hemisphere is essentially different. A circumpolar low is flanked equatorward by high pressure systems in both composites. In the single-jet composite (Fig. 6.11 a), the circumpolar low has a wavenumber 4 structure. A broad trough extends from Tasmania to the mid-Pacific. The pressure gradient on the western flank of the trough over Tasmania is collocated with a region of enhanced DL jet frequency (e.g. Fig. 6.9 a). Although there is no DL jet pattern in Fig. 6.9 (a) collocated with the pressure gradient at the equatorward side of the trough, this former might participate to the presence of the DL jet pattern located at around 25°N east of Australia in Fig. 6.9 (a). The mean sea-level pressure displays a wavenumber 3 structure in the double-jet composite (Fig. 6.11 b). An intense pressure gradient extends around 50°S from south of Australia to the mid-Pacific and is collocated with a region of enhanced DL jet event frequency in Fig. 6.9 (b). Interestingly, no strong high pressure system is found over New Zealand as Fig. 6.6 (f) could have suggested.

Differences in the distribution of the surface cyclone frequencies between both jet configurations are particularly found between Tasmania and the mid-Pacific. In the single-jet composite (Fig. 6.11 c), enhanced frequencies are found from south of Tasmania to the western New Zealand sea shores and in the mid-Pacific north of 60°S. However, no surface cyclones tend to be found equatorward of 30°S in this region. In the double-jet composite (Fig. 6.11 d), a large area south and southeast of New Zealand is characterized by very low cyclone frequencies. The cyclone frequency pattern extends nearly to 20°S in the western and mid-Pacific.

To summarize, differences in the structure and magnitude of mean-sea-level pressure patterns and surface cyclone frequencies have been found between each type of jet configuration and between both hemispheres. In the Northern Hemisphere, single-jet (double-jet) composites show dipoles of opposites synoptic features over the Atlantic: a weak (deep) low pressure system west of Ireland (between southeastern Greenland and northern Scandinavia) and a weak (strong) high pressure system zonally extending from Florida to Egypt (centered over western Europe). A weak (deep) low pressure system is also visible in the Pacific in the single-jet (double-jet) composite. This suggests a variability mode linking the occurrence of single jets in the Atlantic to relatively weak low pressure systems poleward of 40°N in the Atlantic and in the Pacific. Conversely in the Southern Hemisphere, the mean sea-level pressure features a circumpolar low with a wavenumber 4 (3) in the single-jet (double-jet) composite. Surface cyclone frequency patterns appear to be zonal and centered around 40°N over the Atlantic in the single-jet composites in the Northern Hemisphere. In the double-jet composite, the cyclone frequency takes a southwest-to-northeast orientation from the Labrador Sea to the Barents Sea. In the Southern Hemisphere, the cyclone frequency pattern displays enhanced values southwest of New Zealand in the single-jet composite. This could also suggest a re-orientation of the storm track linked to the equatorward displacement of the DL jet mentioned in the former Section and mentioned by Bals-Elsholz et al. (2001) for the polar-front jet.

6.3.4 Further Remarks

In Sections 6.3.2 and 6.3.3, major differences have been put in evidence in features at the tropopause level and baroclinic characteristics between single- and double-jet composites. In the present Section, some further characteristic features of single and double jets are briefly discussed.

Interannual Variability of Single and Double Jets

The results of Section 4.3 suggest that the jet pattern south of New Zealand is quasi-absent in the negative phase of the Antarctic Oscillation (AAO). This fact could be indicative of a link between the jet configuration and the phase of the AAO. It is then interesting to compare the number of single- and double-jet occurrences with the flow variability modes of the North Atlantic Oscillation (NAO) and Arctic Oscillation (AO) in the Northern Hemisphere and of the Antarctic Oscillation (AAO) in the Southern Hemisphere. To this end, an analysis is made in which the single- and double-jet time instances are distributed according to winter monthly indices of the NAO, AO and AAO. The monthly indices used in this study are the same ones used in Section 4.3.
For the NAO, more than 70% of the single-jet events occurd in months with negative indices (Fig. 6.12 a). However, if one defines intense negative (positive) phases as phases having an index smaller than or equal to -1 (larger than or equal to 1), this fraction drops to 18%. Double-jet events (Fig. 6.12 b) occur mainly in months with positive indices (79%) while intense positive-defined months gather 39% of the events.
For the AO, 73% and 45% of the single-jet events are found during months of negative and intense negative phase, respectively. 63% of the double-jet event occur during months of positive phase and 45% during intense positive phase months.
As expected, single-jet events in the Southern Hemisphere tend to be related

to negative AAO phases with 71% of the events occurring during negative phase months. 53% of the events are still found in months with intense negative phases. Most of the double-jet events occur during months of positive phase (62%) and the proportion drops to 28% when considering months of intense positive phases This analysis suggests that single jets (double jets) tend to occur during months

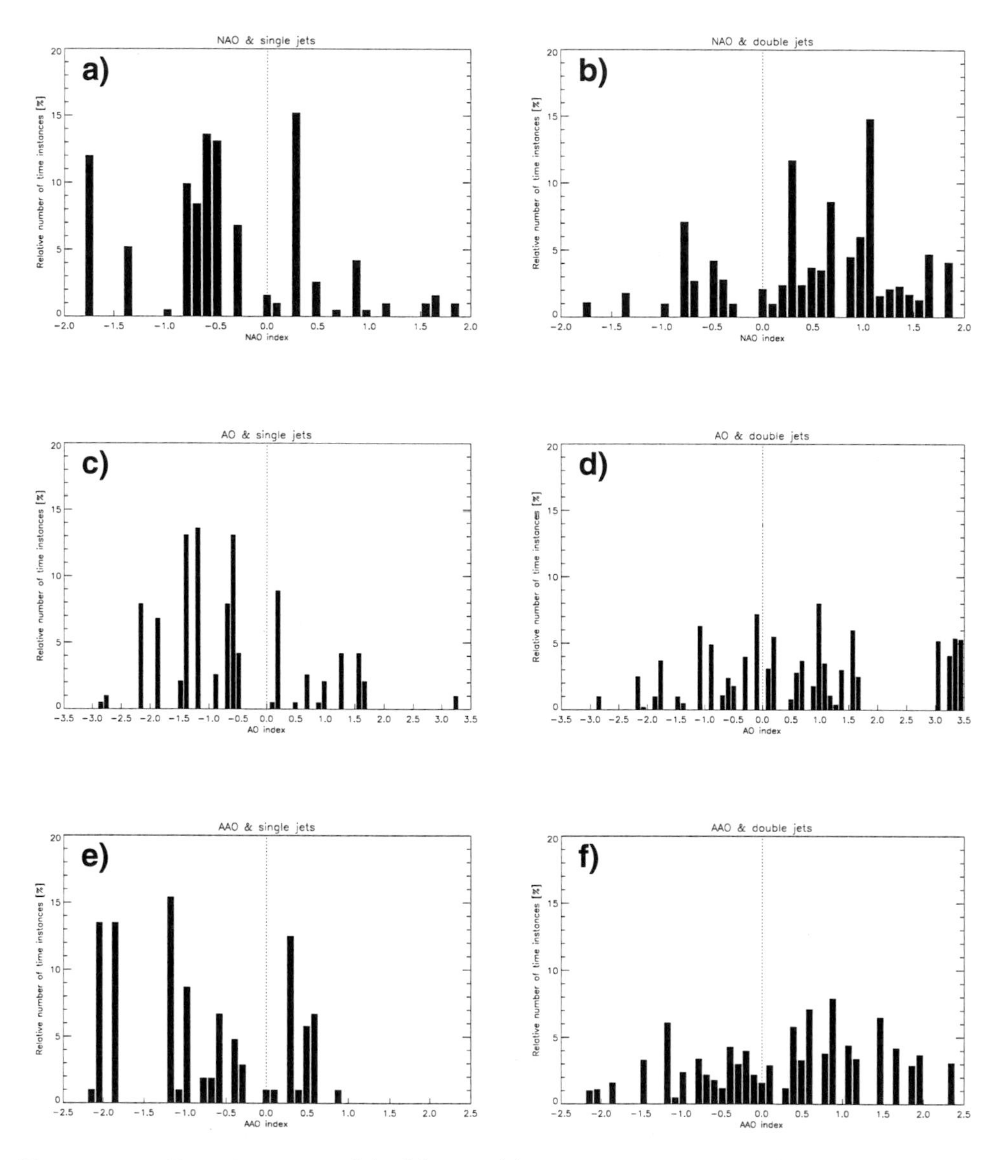

Figure 6.12: Distribution of (a), (c) and (e) the single-jet events (normalized by the total number of events) and (b), (d) and (f) the double-jet events relatively to global flow variability indices (discretized every 0.1). (a) and (b) are for the NAO, (c) and (d) for the AO and (e) and (f) for the AAO.

characterized by negative (positive) indices of the NAO, AO and AAO. However, the results for intense negative and positive phases show quite low proportions of events in both hemispheres, except for the negative phase of the AAO that still gathers 53% of the single-jet events. Clearly, further studies are needed in order to better determine a possible link between the NAO, AO and AAO and the occurrence of single and double jets in the Atlantic and over the western Southern Pacific, respectively. The use of daily indices would also help to refine the analysis.

	Single-Jet Events				Double-Jet Events			
Phase	−−	−	+	++	−−	−	+	++
NAO	18	70	30	5	3	21	79	39
AO	45	73	27	14	16	37	63	45
AAO	53	71	29	0	14	38	62	28

Table 6.1: Respective distributions in % of the single-jet and of the double-jet events following the monthly indices of the North Atlantic Oscillation (NAO), Arctic Oscillation (AO) and of the Antarctic Oscillation (AAO). The −− and ++ columns correspond to intense negative (index smaller than or equal to -1) and positive (index larger than or equal to 1) phases, respectively. The − and + columns correspond to the negative and positive phases, respectively.

Durations of Single and Double Jets

An assessment and classification is made of the duration of single- and double-jet episodes. A large amount of single- and double-jet events have been found that last less than 6 hours. Single-jet episodes are the best representations of this fact with more than 50% of the episodes lasting less than 6 hours in both hemispheres while 47% and 33% of the double-jet episodes are affected in the Northern and Southern Hemisphere, respectively. These short-lasting episodes are not taken into account for this analysis.
In the Northern Hemisphere (Fig. C.1 a in the Appendix), the relative frequency of multiple-jet episodes is inversely proportional to the duration. The mean durations are 19 hours and 21 hours for the single jets and double jets, respectively. Episodes of both types of flow configuration are relatively short-lived as nearly 70% of them last less than 24 hours.
In the Southern Hemisphere (Fig. C.1 b in the Appendix), double-jet episodes have a much longer mean duration than in the Northern Hemisphere (32 hours). The mean duration of the single-jet episodes is similar to the boreal hemisphere (19 hours). However, most of the single-jet episodes (91%) last less than 24 hours.

Conversely, nearly 60% of the double-jet episodes duration do not exceed one day. Thus, in both hemispheres, single-jet episodes are shorter-lived than double-jets. In the Southern Hemisphere, this difference is more pronounced. Single-jet episodes have the same mean duration in both hemispheres. However, most of the single-jet episodes in the Southern Hemisphere do not last more than a day, while the fraction of the episodes lasting more than a day in the Northern Hemisphere is larger (more than 30%). More long-lasting double-jet episodes are found in the Southern Hemisphere. These results show a temporal difference of single and double jets between both hemispheres. Both single-jet and double-jet episodes can last more than a day in the Northern Hemisphere. In the Southern Hemisphere, single-jet episodes lasting more than a day have to be considered as extreme cases, as most episodes do not exceed 24 hours while the double-jet configuration can last for much longer periods.

PV gradients on isentropic surfaces

Further insights into the differences between the single-jet and double-jet states can be gained from the study of PV gradients on isentropic surfaces (e.g. Schwierz et al. 2004). Jet streams are here recognizable as regions of enhanced PV gradients where the isentropic surface intersects the tropopause break. PV contours and gradients are investigated on the 320 K, 340 K and 440 K isentropic surfaces. The first two isentropic surfaces interesect the polar and subtropical tropopause breaks, respectively (e.g. Shapiro et al. 1999). The 440 K isentropic surface is located in the stratosphere between 50 and 100 hPa at midlatitudes and sheds light on the polar vortex.
In the Northern Hemisphere, the single-jet composite is characterized by a zonal band of enhanced PV gradient extending from eastern America to western Europe on both the 320 K (Fig. 6.13 a) and 340 K (Fig. 6.13 c) isentropic surfaces. Large values of the PV gradient are found over France and Spain. Conversely, there is a smaller PV gradient over France and Spain in the double-jet configuration, while large values are found from Nova Scotia to Scandinavia on the 320 K surface (Fig. 6.13 b). PV contours display a ridge structure over western Europe and the PV gradient is discontinuous in this region. On the 340 K surface (Fig. 6.13 d), an intense PV gradient extends from the Canary Islands to the mid-Pacific northwest of Hawaii and over the United States. No PV gradient is present then over the European sector and the PV contours adopt a northwest-to-southeast orientation over the mid-Atlantic.
The single-jet composite in the Southern Hemisphere is characterized by a trough-like pattern of the PV contours on the 320 K surface (Fig. 6.14 a) south of Tasmania and New Zealand. Enhanced PV gradients extend from southwestern Australia to the eastern Indian Ocean some 20° poleward. The double-jet composite displays two zones of enhaced PV gradient south of 20°S on the 320 K

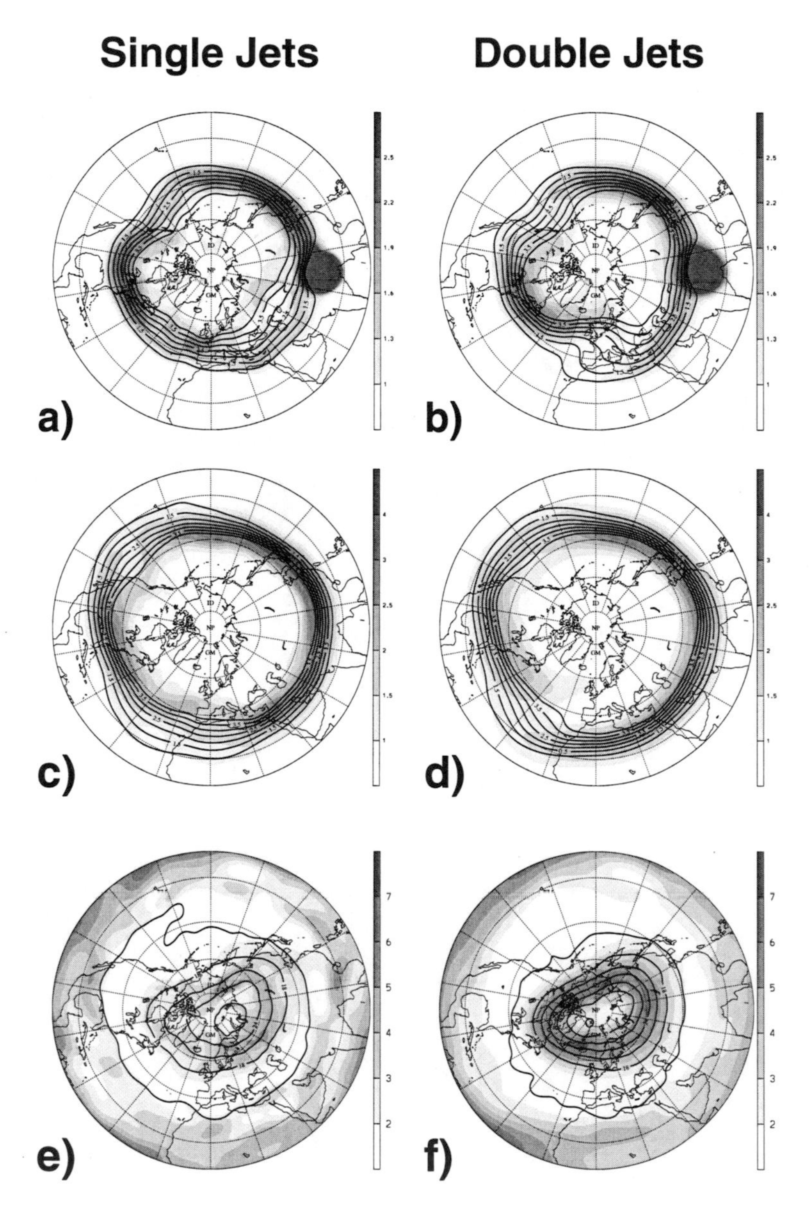

Figure 6.13: Potential vorticity gradients (grey shaded, units 10^{-6} pvu·m^{-1}) and contours (black solid lines, units pvu) in the Northern Hemisphere on the (a,b) 320 K, (c, d) 340 K and (e, f) 440 K isentropic surfaces for (a, c, e) the single-jet and (b, d, f) the double-jet composite. Note that the scales for the PV gradient differ from one isentropic surface to another. Regions of high PV gradient northeast of India in (a) and (b) are caused by the intersection of the 320 K isentropic surface with orography.

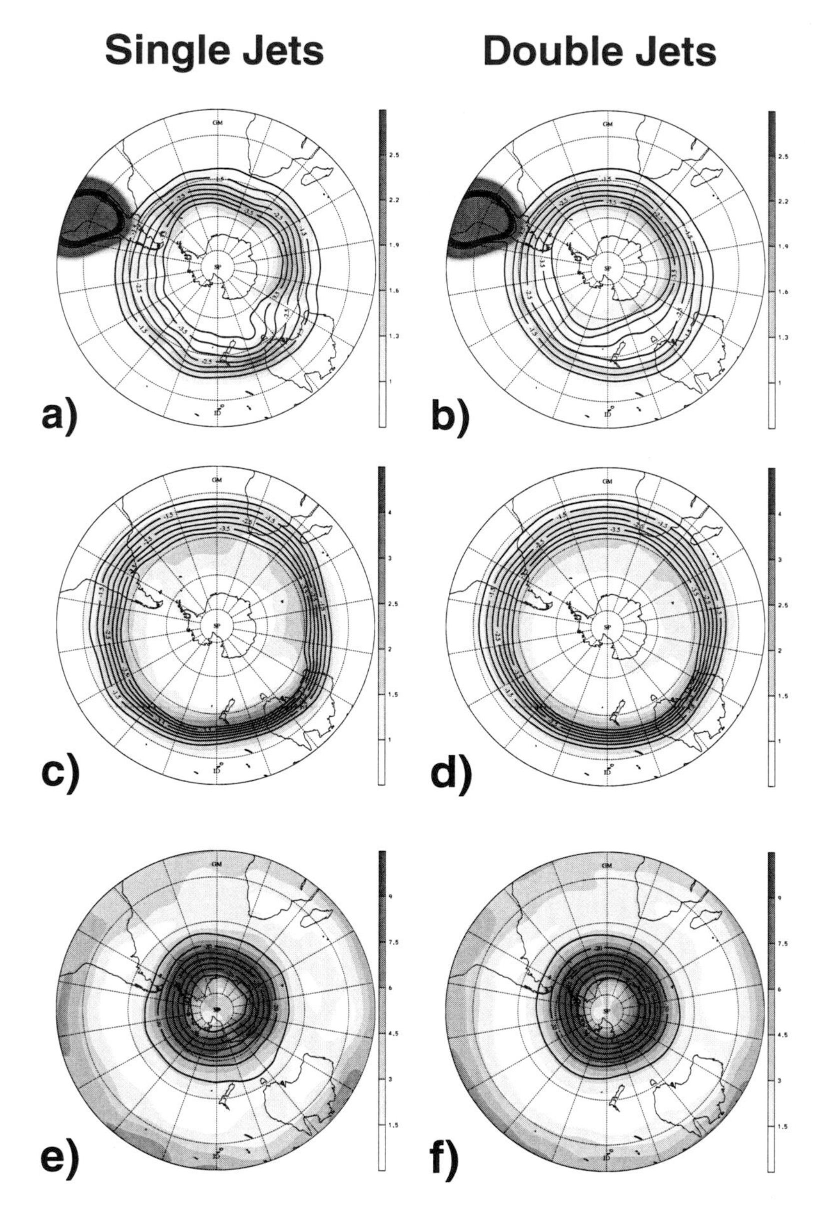

Figure 6.14: Potential vorticity gradients (grey shaded, units 10^{-6} pvu·m^{-1}) and contours (black solid lines, units pvu) in the Southern Hemisphere on the (a, b) 320 K, (c, d) 340 K and (e, f) 440 K isentropic surfaces for (a, c, e) the single-jet and (b, d, f) the double-jet composite. Note that the scales for the PV gradient differ from one isentropic surface to another. Regions of high PV gradient over western South America in (a) and (b) are caused by the intersection of the 320 K isentropic surface with orography.

surface meridionally separated by a PV contour having a slight ridge structure (Fig. 6.14 b): one is located around 30°S and extends from southwestern Australia to east of New Zealand and the other one around 60°S. Note that the results on the 320 K isentropic surface emphasize the advantage of the SL/DL partition introduced in Chapter 5 upon the polar-front/subtropical jet nomenclature. As a matter of fact, as the 320 K intersects the polar tropopause according to Shapiro et al. (1999), this would mean that the polar tropopause (i.e. the polar jet) would be located around 30°S in both composites, i.e. in the region where the subtropical jet is supposed to be found. No significant difference appears on the 340 K surface between the single- and double-jet composites, except for a slight increase of the PV gradient from the eastern Indian Ocean to the western Pacific (Fig. 6.14 c and d).

The stratospheric polar vortex is located at high latitudes and is characterized by a region of high PV and a strong jet associated with steep PV gradients at its edge (e.g. Schoeberl et al. 1992). These vortices exhibit day-to-day variabilities (Waugh and Randel 1999) and intrusions into or extrusions out of the vortices are possible (Plumb et al. 1994). The variability is larger in the Northern Hemisphere where upward propagating waves cause stratospheric sudden warming episodes (Davies 1981), decelerating the stratospheric jet (Zhou et al. 2000). However the connections between stratospheric and tropospheric flows are not completely understood. Some studies point to a troposphere-to-stratosphere influence (e.g. Ambaum and Hoskins 2002) and others to stratospheric precursors of tropospheric flow regimes (Baldwin and Dunkerton 2001). The single- and double-jet composites of the PV gradients and contours on the 440 K isentropic surface shed light on the properties of the winter lower stratospheric Arctic and Antarctic vortices in single-jet and double-jet flow configurations.

In the Northern Hemisphere, the double-jet composite displays a compact vortex centered near the North Pole and delimited by intense PV gradients, especially from Baffin Island to Iceland. Conversely, the PV gradients are weak in the single-jet composite, probably indicating a weaker polar vortex. This outlines the simultaneous presence of a strong polar vortex (in particular its North Atlantic section) and of a southwest-to-northeast oriented underlying PV gradient and associated jet in the double-jet composite.

In the Southern Hemisphere, both jet composites exhibit a compact strong polar vortex defined by intense PV gradients poleward of 60°S. However, the PV gradient in the single-jet composite displays slightly smaller values south of Tasmania together with a slight trough structure of the PV contours.

6.4 Summary

In this Chapter, characteristics of winter single- and double-jet events have been inferred from the event-based jet-stream climatology of Chapter 4.
First, the zonal frequency distribution of single jets and double jets has been studied. It has been found that single-jet events are more frequent in the Northern Hemisphere than in the Southern Hemisphere. Each type of flow configuration favours particular regions and their distribution is quite variable from one hemisphere to another. Two regions with a clear prevalence of the single-jet configuration have been found in the Northern Hemisphere (western Pacific and western Atlantic), while double jets tend to occur more frequently over the mid-Atlantic-European sector. In this region, the double-jet pattern consists climatologically of two distinct jets located poleward and southward of 40°N, respectively. In the Southern Hemisphere, double-jet configurations prevail clearly from the eastern Indian Ocean to the mid-Pacific and both types of configuration are nearly equally found in the rest of the hemisphere. The two jet patterns are found climatologically in the 20°S-40°S belt and poleward of 45°S, respectively.
Second, temporal and spatial coherent single-jet and double-jet episodes have been searched for in the regions with enhanced double-jet event frequencies and composites have been calculated. It has been found that single-jet events in the Northern Hemisphere and Southern Hemisphere have different flow properties. In the Northern Hemisphere, single-jet events are on average characterized by a jet feature meridionally located between the climatological positions of the two jets of the double-jet composite (around 40°N). In the Southern Hemisphere, the southernmost jet tends to disappear south of Australia and New Zealand while the meridional position of the northernmost jet remains the same in both the single- and double-jet configuration. Results suggest that the poleward jet might actually migrate equatorward in the single-jet composite due to the outbreak of cold Antarctic air south of New Zealand.
Single-jet and double-jet composites have been studied in the shallow-layer baroclinicity (SL) and deep-layer baroclinic (DL) perspective introduced in Chapter 5. Results show that the jet in the single-jet composite in the Northern Hemisphere has DL characteristics and SL characteristics in the Southern Hemisphere. Differences in the positions of the storm track in relation to the meridional jet configuration have also been inferred. In the Northern Hemisphere, the single-jet configuration is associated with a zonal Atlantic storm track and storms are likely to affect the west coast of the British Isles and northwestern France while in the double-jet configuration, storms take a southwest-to-northeast path and affect primarily Scandinavia and the north of Europe. Slight differences have also been found in the Pacific. In the Southern Hemisphere, more storms affect Tasmania and New Zealand in the single-jet configuration.
Further characteristics of single and double jets have been briefly reviewed. Thus,

some elements regarding the connection between the occurrence of single and double jets and the North Atlantic Oscillation, Arctic Oscillation and Antarctic Oscillation have been explored. To this end, the dates of the single- and double-jet events have been compared to monthly indices of the variability modes. Results suggest that single-jet events are mostly found in months characterized by a negative phase of the variability, while double-jet events tend to occur during positive phase-defined months. The same study with daily indices might enable a further refinement of this statement. Some insights to the durations of multiple jets have also been given. The mean durations of single-jet episodes are on average shorter than double-jet episodes. This is particularly true in the Southern Hemisphere where most of the determined single-jet episodes last less than a day. However, single-jet episodes of more than a day are more common in the Northern Hemisphere than in the Southern Hemisphere.
Finally, the PV contours and gradients have been investigated on the 320 K, 340 K and 440 K isentropic surfaces. Results for the Southern Hemisphere on the 320 K isentropic surface illustrate the greater consistency of the SL/DL classification compared to the classical subtropical/polar-front jet nomenclature. Besides, results on the 440 K isentropic surface suggest the occurrence of a weak polar vortex for the single-jet composites, particularly in the Northern Hemisphere.

Part II

Relative Roles of the Atlantic and Mediterranean as Moisture Sources in Two Heavy Precipitation Episodes on the Alpine Southside

Chapter 7

Introductory Remarks

A particular form of meteorological extreme events is considered in this Part of the thesis. The focus is set on heavy precipitation episodes on the Alpine southside. After some general considerations relative to heavy rain hazards and floods, synoptic and orographic characteristics that favour extreme rainstorms on the Alpine southside are briefly reviewed. A summary of the motivations and objectives of this part concludes the Chapter.

7.1 General Considerations

Among the various forms of natural hazard, floods are by far the type that affects the most people and causes the most damages (WMO 2004). They remain however difficult to predict. Floods are mainly linked to precipitation. Doswell et al. (1996) stated that the total amount of precipitation produced in an event was proportional to the rainfall rate and the duration. Floods can therefore arise from extreme rainfall rates over relatively short time periods (and the effect will be all the more devastating when the rainstorm tends to be stationary) or from moderate rainfall rates lasting over relatively long periods. The heavy precipitation episodes studied here are characterized by rainfall amounts that range from severe to extreme and that can last over long periods (from 2 to 5 days).
Mountainous regions such as the Alps are very sensitive to severe rainfall. Most damages caused by heavy rainstorms in the Alps are due to the combination of two factors. First, rivers that abruptly overflow owing to the sudden increase of their discharge after an episode of heavy precipitation constitute the hydrological factor. Second, the favoring effect of the slopes of the mountains for the onset of landslides constitutes the 'gravity' factor. The association of both factors can have dramatic consequences in terms of the extent of the damages and of human

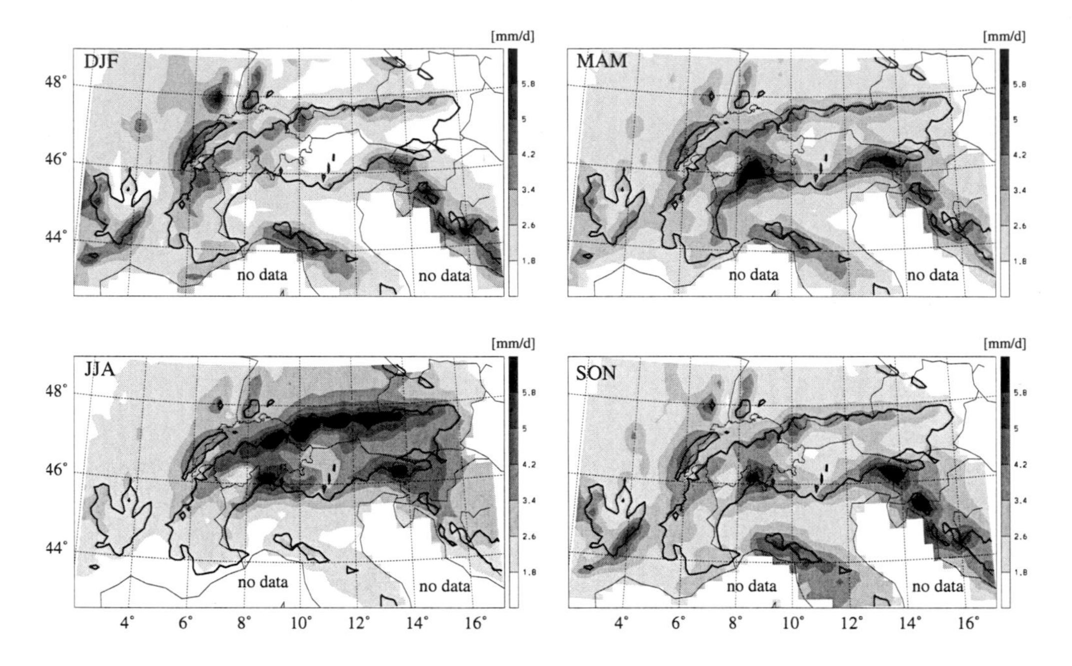

Figure 7.1: Mean seasonal precipitation in the Alpine region for the period 1971-1990 for winter (DJF), spring (MAM), summer (JJA) and autumn (SON). From Frei and Schär (1998).

casualties. For instance the material costs of the September 1993 and October 2000 flood events that will be investigated in this study exceeded 550 millions and 440 millions Euros (BWG 2002), respectively. Moreover, they claimed 2 and 16 lives, respectively.

Frei and Schär (1998) compiled a rain climatology based on observations over the 20-year period from 1971 to 1990 for the Alpine region. They found the largest mean precipitation to occur on the southern slopes of the Alps and the Massif Central, especially during autumn and spring (Fig. 7.1). The fact that the Alpine southside is so severely affected by precipitation can be understood when two factors are taken into account. First, the Mediterranean lies directly south of both mountain ranges and can serve as a moisture source when a warm southerly flow exists. Second, the mountain ranges themselves can enhance the lifting of the impinging air masses that in turn destabilizes the atmosphere and favours convection. Frei and Schär also pointed out the concentration of heavy precipitation into relatively short and intense events. The successful forecast of these extreme events in mountainous regions has therefore become a challenge for modern meteorology. However, these episodes are extremely complex to diagnose because they involve a series of phenomena ranging from the meso-α scale (fronts) to the micro-ϵ scale (precipitation and cloud particles, e.g. Davies 1999,

Fig. 1.4). Nevertheless, the use of numerical models and of weather data sets have led to progress in the understanding of these events. Moreover, large measurement campaigns have been elaborated in the Alpine regions in order to collect a large amount of data to enable a better understanding of the different involved phenomena: ALPEX (ALPine EXperiment, Küttner and co-authors 1980, Davies and Pichler 1990) and more recently MAP (Mesoscale Alpine Program, Binder and Schär 1996).

7.2 Synoptic Characteristics

In recent years, many episodes of flash floods have affected the southern slopes of the Alps (Vaison-la-Romaine, September 1992; Brig, September 1993; Piedmont, November 1994; Gondo, October 2000; Schlans, November 2002) and caused extensive damages and human casualties. Classically, three synoptic ingredients are needed for southern Alpine rainstorms to occur: (i) a large supply of moisture in the lower troposphere towards the Alpine southside, (ii) a vertical stratification favourable for deep convection and (iii) a mechanism to trigger and maintain deep convection. These features occur when a deep trough develops over the eastern Atlantic-western European sector and extends sufficiently southward so that the associated frontal structure can reach the western Mediterranean. The triggered low-level jet can then advect in a southwesterly to southerly flux warm moisture-laden air masses towards the Alpine region. The orography tends then to provoke the lifting of the air masses, which in turn favours convection. Note that if the front is advected eastward relatively slowly, this situation can continue several days.
In their study of four different heavy precipitation episodes on the Alpine southside, Massacand et al. (1998) found the recurrent presence of an upper-level precursor in the form of a meridionally elongated narrow intrusion of stratospheric air from the British Isles to the western Mediterranean. The presence of this upper-level positive potential vorticity anomaly over western Europe can account for the classical synoptic features by enhancing the southerly component of the flow toward the Alpine region and reducing the static stability on its eastern flank. Further research has shown the relevance of mesoscale structures within the PV streamer for a quantitative improvement of the forecasted precipitation on the Alpine southside (Fehlmann et al. 2000 and Fehlmann and Quadri 2000).

7.3 Orographic Effects

Frei and Schär (1998) found in their rain climatology for the Alpine region an inhomogeneous distribution of the rainfall amounts, some regions like the Lago Maggiore displaying in average larger precipitation amounts than their surroundings (Fig. 7.1). This is due to orographic mechanisms that can locally concentrate and enhance precipitation and cause flash floods.

The study of flow past orography is an important topic of research in geophysical fluid dynamics. Studies using numerical models have revealed a large spectrum of possible flow configurations sensitive to different parameters of the impinging flow (velocity, static stability) and of the dimensions and shapes of the mountain range (e.g. Smith 1989, Gheusi 2001 and references therein, Gaberšek 2002). These parameters control among others two possible flow configurations through the inverse Froude number (e.g. Davies 1997): the 'flow over' and the 'flow around' regimes. A high flow speed, low mountain height and low vertical stratification will favour the flow over regime. In this case, air is forced to rise by the orography. If moist air is impinging upon the mountain, condensation and orographic precipitation will occur (Houze 1993, Section 12.4). Conversely, a high mountain, a low flow speed and a strong vertical stratification will tend to favour the flow around regime. In this case, the air flow is forced to circumvent the mountain[1].

In the case of heavy precipitation episodes on the Alpine southside, Buzzi *et al.* (1998) put in evidence the relevance of condensation for modifying the flow on the upstream side of the mountain range and for the interaction between the flow and the orography. In effect, the release of latent heat of condensation favours the 'flow over' regime to the detriment of the 'flow around' by destabilizing the vertical stratification of the lower troposphere (Buzzi and Foschini 2000). In idealized studies of the 'flow around' regime, Pierrehumbert and Wyman (1985) found an easterly jet (in the Northern Hemisphere) along the upwind mountain slope (the *barrier wind*, Fig. 7.2) due to the leftward deflection (toward the lower pressure) of the slowing flow approaching the obstacle when the Coriolis parameter f is taken into account. This asymmetric split of the flow was later confirmed by the dry simulations of Schneidereit and Schär (2000). The relevance of the barrier wind for the 1994 Piedmont flood was studied by Rotunno and Ferretti (2001) and for the maximum in the average October precipitation in the region of the Lago Maggiore by Gheusi and Davies (2004). They found that different regions of the Alpine arc were affected by precipitation when the impinging flow's direction was changed from southwesterly to southeasterly.

[1]Note that for large mountains (L>50 km), the inverse Rossby number ($= Lf/U$) has also to be considered for the distinction between the 'flow over' and 'flow around' configurations (Trüb and Davies 1995).

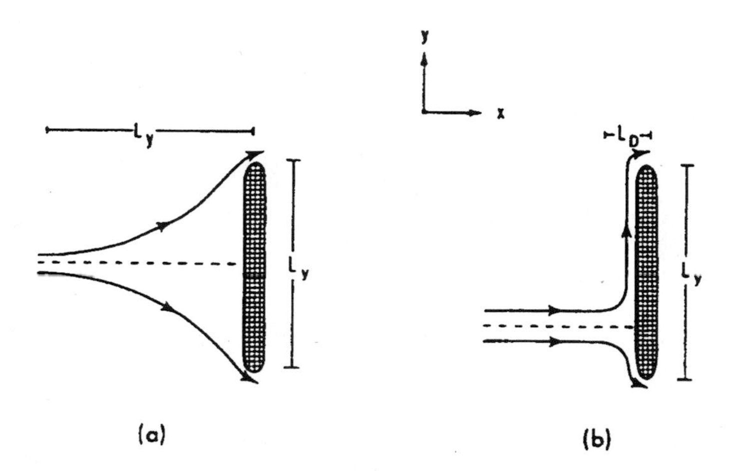

Figure 7.2: Schematic flow pattern upstream of a mountain in the rotating case (Coriolis force taken into account). L_y denotes the long dimension of the mountain, L_D ($= NH/f$) is the characteristic scale of the upstream influence of the mountain. If $L_D >> L_y$ (a), the flow is nondivergent and is not affected by the Coriolis force. If $L_y >> L_D$ (b), the flow becomes subgeostrophic and the Coriolis force creates an along mountain wind, causing the fluid to flow around the mountain to the left. Schematic from Pierrehumbert and Wyman (1985). See text for more details.

7.4 Motivation and Aims of Part II

Earlier studies have pointed out that the Mediterranean is the main moisture supplier in heavy precipitation episodes on the Alpine southside (Massacand et al. 1998). However, the relatively small dimensions of the Mediterranean and the nearby presence of a much larger moisture source (the Atlantic) prompt the investigation of the relative role of the Mediterranean and alternative moisture sources. This interest is further motivated by the 'origins' of 72-hour backward trajectories of air parcels having a relative humidity larger than 80% (i.e. raining condition) in the Alpine region in two heavy precipitation episodes. In the September 1993 flood event, the origins of the backward trajectories started at 00 UTC 23 September (Fig. 7.3 a) and 24 September 1993 (Fig. 7.3 b) are the eastern Mediterranean and the Sahara. However, the situation is quite different in the October 2000 episode, as the origins of the trajectories differ depending on the starting date. The origin of the 72-hour backward trajectories started at 00 UTC 13 October 2000 is clearly the Atlantic (Fig. 7.3 c). Conversely, the backward trajectories started at 00 UTC 15 October 2000 originate from the Mediterranean and south of Algeria (Fig. 7.3 d).
Beside its impacts on the precipitation, several studies have shown the relevance of moisture (through diabatic effects in condensation) for the production of a vertical PV dipole with a positive PV anomaly below and a negative PV anomaly

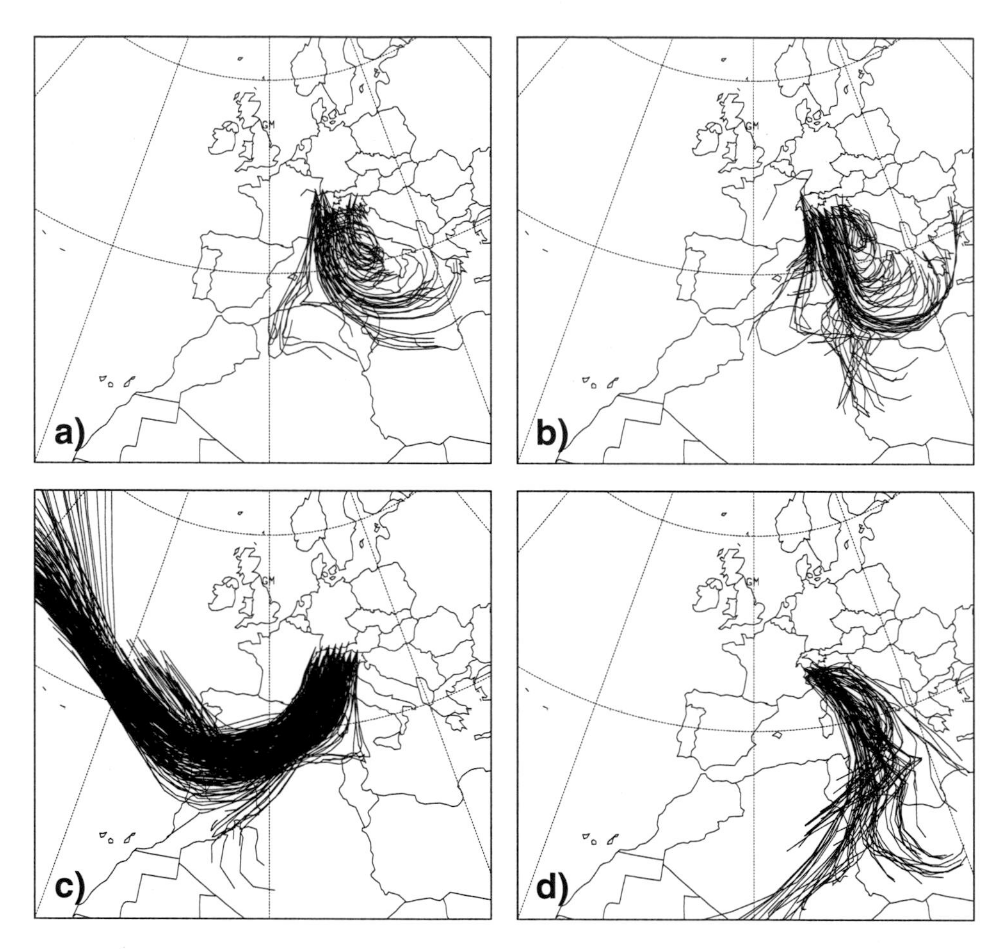

Figure 7.3: 72-hour backward trajectories calculated from the ECMWF analysis data started between 900 hPa and 500 hPa in the Alpine region. Calculations started at (a) 00 UTC 23 September 1993, (b) 00 UTC 24 September 1993, (c) 00 UTC 13 October 2000 and (d) 00 UTC 15 October 2000 for points having a relative humidity larger than 80%.

above the level of condensation (see Section 2.1.3). Frontal structures are meteorological features where the ascent of moist warm air and condensation in the warm conveyor belt (Browning 1990) are favoured. Airflow climatologies reveal the relevance of the eastern seaboards of North America and Asia as origins of warm conveyor belts (Stohl 2001, plate 1; Eckhardt et al. 2004). Recently, studies have tackled the difficult task of isolating the effects of the diabatically produced reduced upper-tropospheric PV (Stoelinga 1996, Pomroy and Thorpe 2000, Stoelinga 2003, Thorpe 2003). Massacand et al. (2001) studied the effects of frontogenesis over the western North Atlantic (and the associated production of reduced upper-tropospheric PV) upon the downstream development of the area of negative PV anomaly and the subsequent downstream development of the PV streamer that led to the September 1993 heavy precipitation episode. By means of idealized numerical simulations with and without cloud-diabatic effects, they established a dependence of the PV streamer's generation upon the preconditioning of the upstream upper-tropospheric flow characteristics by diabatic effects.

Aims of Part II

Many questions related to the role played by the Atlantic and the Mediterranean in Alpine rainstorms still remain unanswered. Figure 7.3 shows that although the Mediterranean can be regarded as a major moisture supplier in heavy precipitation episodes on the southern side of the Alps, there are other possible moisture sources and that their involvement varies from one episode to another. Moreover, the impacts of diabatic heating upon the evolution of the upper-level PV streamer has been studied for a single case through the removal of diabatic heating. This only gives a general idea of the relevance of diabatic processes for the genesis of the PV streamer but does not permit to isolate the relevance of each water basin as moisture provider that will eventually be involved into condensation processes and the release of diabatic heating.
The aim of this Part is to study the impacts of the Mediterranean and of the Atlantic upon two heavy precipitation episodes on the Alpine southside. Using numerical simulations in which the evaporation from the Atlantic and the Mediterranean are suppressed, this study will focus on:

- the respective effects of the Atlantic and the Mediterranean as moisture sources upon the distribution and intensity of rainfall in the Alpine region during two heavy precipitation episodes.

- the identification of potential supplementary moisture sources beside the two water basins.

- the attribution of the effects of the Atlantic and Mediterranean moisture upon the development of the upper-level PV streamer.

In Chapter 8, the methodology used in this study and the two heavy precipitation episodes are presented. Chapter 9 focusses on the individual and combined effects of the Atlantic and the Mediterranean upon the precipitation. The influence of another potential moisture source is explored. In Chapter 10, the impacts of both water basins upon the development of the upper-tropospheric PV streamer are explored.

Chapter 8

Case Description and Methodology

In this Chapter, the two heavy precipitation episodes studied in this Part are presented. A description is given of the methodology and the different types of simulations are presented.

8.1 Case Description

Two episodes of heavy precipitation on the Alpine southside have been selected for this study. These two flood events affected the southeastern regions of the Swiss Cantons of Valais and Ticino, causing extended damages and claiming human lives.

8.1.1 The September 1993 Flood Event (The Brig Event)

Between 22 and 25 September 1993, several synoptic systems resulted in large amounts of rainfall in southern Europe (Fig. D.1). In the Simplon pass area (southern Switzerland), 24-hour accumulated precipitation values in excess of 120 mm and 220 mm were recorded on the 23th and 24th, respectively (Benoit and Desgagné 1993). Specifically, the Swiss town of Brig, located directly north of the Alpine divide was devastated by a flood. For this study, the September 1993 flood event will also be referred to as the Brig event.
The synoptic situation at the beginning of the period is characterized by a 500-hPa low located southeast of Iceland at 12 UTC 21 September (Fig. 8.1 a). A trough extends southward from Iceland to Portugal and a ridge is present over

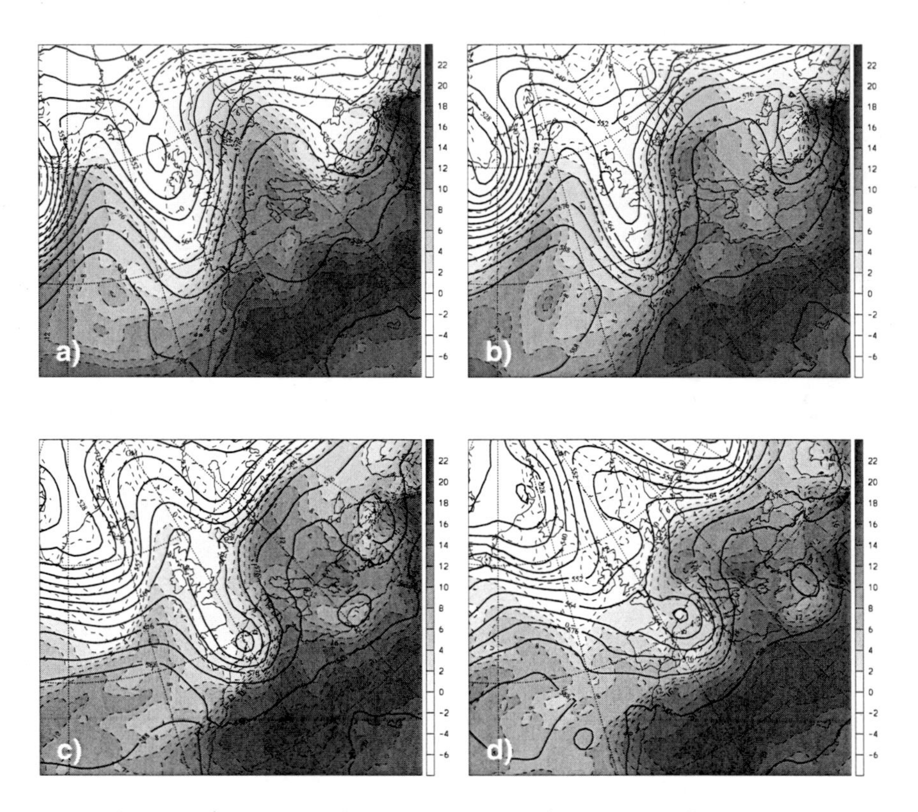

Figure 8.1: Synoptic evolution for the Brig case (from the ECMWF analysis data). 500 hPa geopotential (solid contours in dam, from 516 to 588 dam, every 6 dam) and temperature at 800 hPa (shaded and dashed contours every 2 K) at (a) 12 UTC 21 September 1993, (b) 12 UTC 22 September 1993, (c) 12 UTC 23 September 1993 and (d) 12 UTC 24 September 1993.

central Europe. A cold front extends at 800 hPa from southern Portugal to almost the North Sea. At 12 UTC 22 September (Fig. 8.1 b) the trough axis extends from the United Kingdom to Gibraltar. The 800-hPa cold front strengthens, extends from southwest of Gibraltar to Denmark and is about to reach the western Mediterranean. 24 hours later (Fig. 8.1 c), a cut-off low appears over the coasts of Catalonia in the 500-hPa geopotential and the ridge is shifted eastward to eastern Europe. The cold front extends from northern Algeria over the Alpine region to the Baltic Sea. Finally at 12 UTC 24 September, the cut-off low has moved northeastward to southwestern France and the trough axis has taken a northwest-to-southeast orientation (Fig. 8.1 d). The front extends from Tunisia to the Alpine region.

The upper-level synoptic situation at the beginning of the episode is characterized by a meridionally elongated and narrow positive PV anomaly extending west of the British Isles (Fig. 8.2 a). At 12 UTC 22 September, the PV streamer has

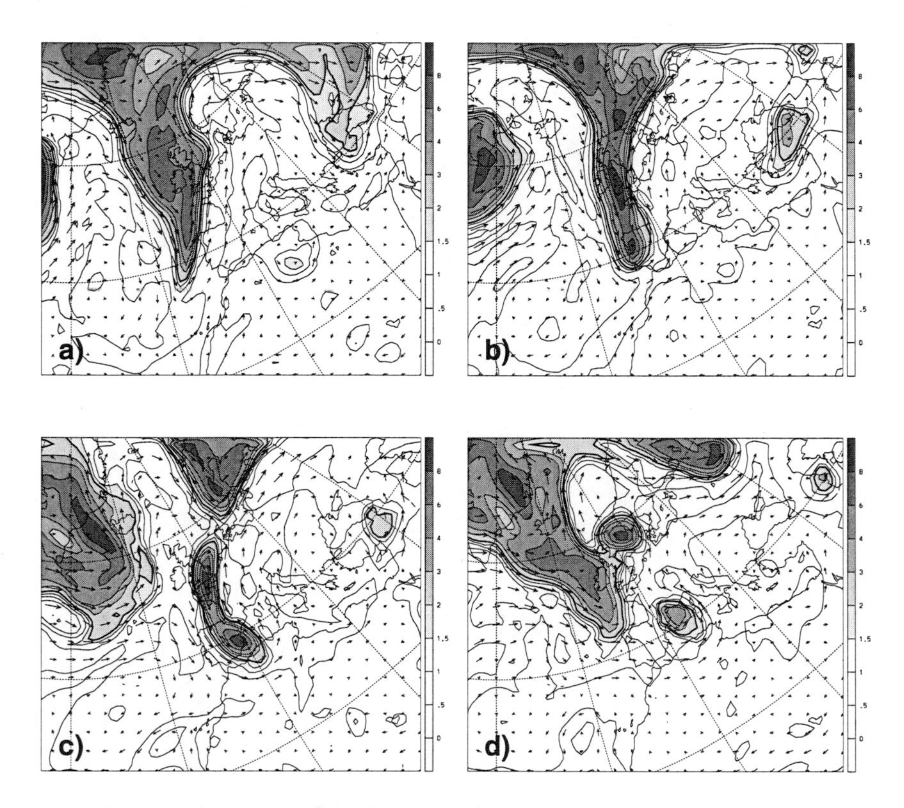

Figure 8.2: Potential vorticity (shaded in pvu) and wind vectors on the 320 K isentropic surface diagnosed from the ECMWF analysis data for the Brig case at (a) 12 UTC 21 September 1993, (b) 12 UTC 22 September 1993, (c) 12 UTC 23 September 1993 and (d) 12 UTC 24 September 1993. The 2-pvu line is outlined by a bold black line.

reached western France and extends to Spain (Fig. 8.2 b). 24 hours later (Fig. 8.2 c), the streamer has cut-off from the main PV reservoir. The southern part of the streamer has reached the Mediterranean, while the northern part has slightly shifted eastward. At 12 UTC 24 September, the stratospheric cut-off has split into two parts (Fig. 8.2 d). The southern part has settled over the Balearic Islands, while the northern part has moved northward toward Scandinavia where it is about to merge with a larger PV anomaly located over the eastern Atlantic. Note that throughout the episode, the flow over the Alps on the 320 K isentropic surface has been mainly southerly to southeasterly.

8.1.2 The October 2000 Flood Event (The Gondo Event)

The October 2000 heavy precipitation episode was exceptional owing to its long duration (from 11 October to 16 October). Beside its unusual length, two spe-

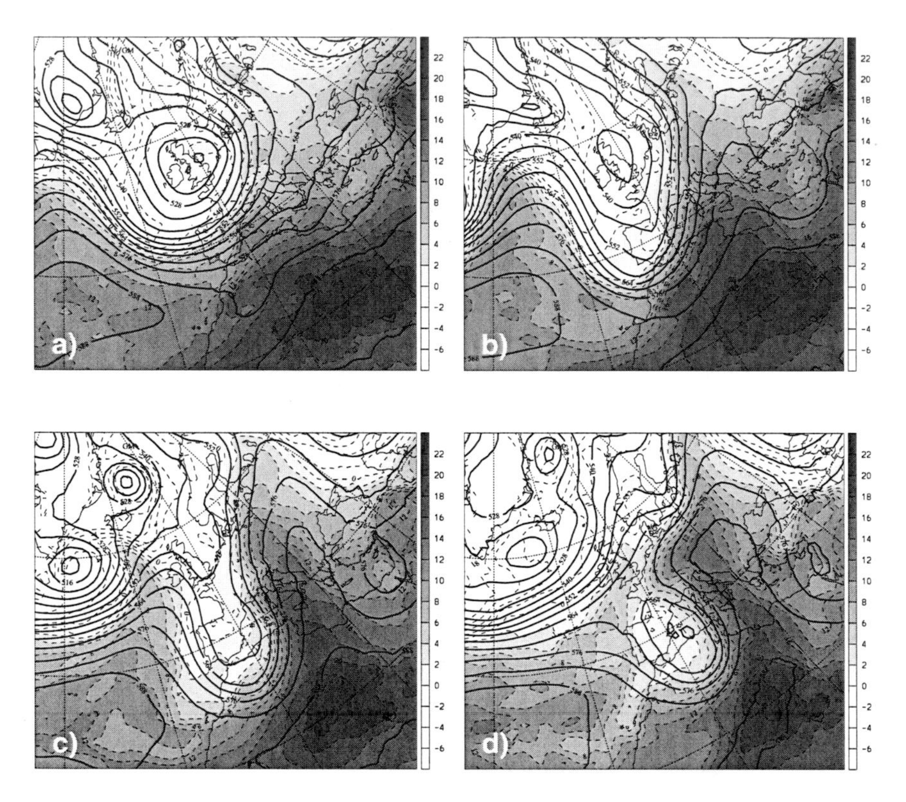

Figure 8.3: Synoptic evolution for the Gondo case. 500 hPa geopotential (solid contours in dam, from 516 to 588 dam, every 6 dam) and temperature at 800 hPa (shaded and dashed contours every 2 K) at (a) 12 UTC 11 October 2000, (b) 12 UTC 12 October 2000, (c) 12 UTC 13 October 2000 and (d) 12 UTC 14 October 2000.

cial features characterize this episode. First, the currents over the Alps gradually turned from southwesterly on 12 October to southerly on 13 October and southeasterly afterward. Second, the snowline rose from 2000 meters on 11 October to nearly 3000 meters between the 13th and the 15th (BWG 2002, Chapter 2). Heavy precipitation fell in the southern Alpine region during the episode (Fig. D.2), especially in the Simplon pass region whose daily accumulated values exceeded 240 mm on both the 13th and 14th. The rapid melting of the snow below 3000 meters combined with heavy rains dramatically increased run-off and provoked extended damages due to flooding and mud slides. Among other damages, the Swiss village of Gondo in the Simplon pass area was partially destroyed by mud slides. Note that for this study, the October 2000 flood event will also be referred to as the Gondo event.
The synoptic situation at the beginning of the episode is dominated by a very deep low at 500 hPa that has settled over the British Isles (Fig. 8.3 a). The associated cold front at 800 hPa is nearly zonal between the eastern Atlantic and Spain and

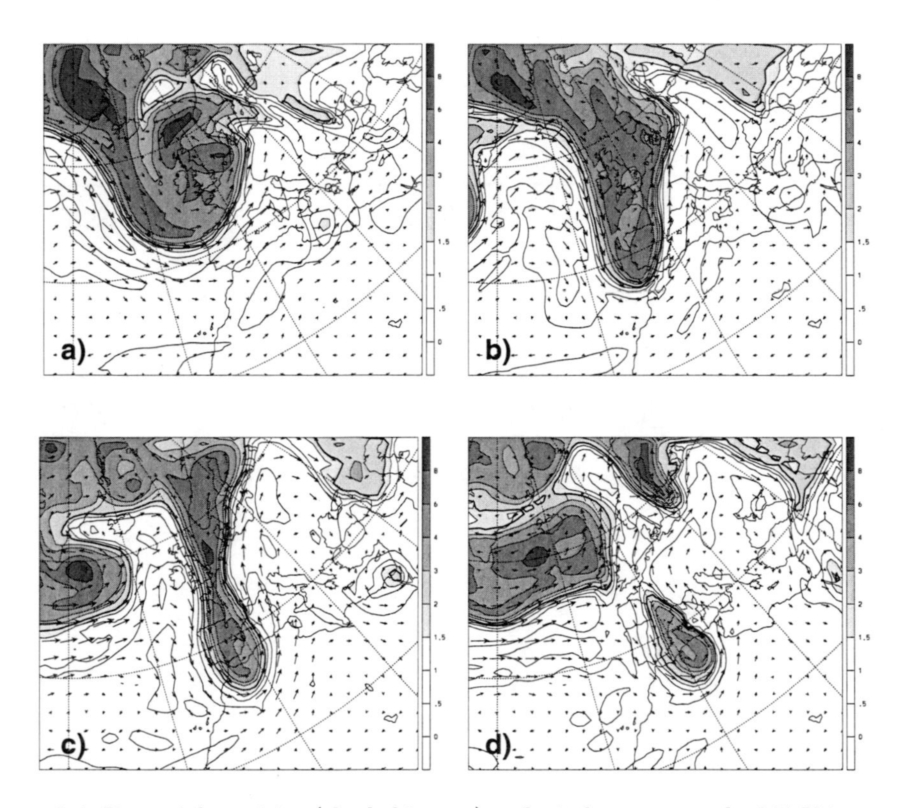

Figure 8.4: Potential vorticity (shaded in pvu) and wind vectors on the 320 K isentropic surface diagnosed from the ECMWF analysis data for the Gondo case at (a) 12 UTC 11 October 2000, (b) 12 UTC 12 October 2000, (c) 12 UTC 13 October 2000 and (d) 12 UTC 14 October 2000. The 2-pvu line is outlined by a bold black line.

westsouthwesterly in the Alpine region. At 12 UTC 12 October, the low pressure system has weakened and the associated trough has extended southward to the coasts of Morocco (Fig. 8.3 b). A ridge is also forming over eastern Europe. The cold front has taken a southwest-to-northeast orientation and extends from the western coasts of Morocco over the Alpine region to the Baltic Sea. At 12 UTC 13 October (Fig. 8.3 c), the trough-ridge structure has further developed, the trough southern tip reaching central Algeria while the ridge extends to Russia, east of the Gulf of Finland. The associated low-level front at the eastern flank of the trough has strengthened and extends from central Algeria over the Alpine region to the Baltic Sea. At 12 UTC 14 October (Fig 8.3 d), a 500-hPa cut-off low has formed over the Balearic Islands. The trough axis has taken a northwest-to-southeast direction. The ridge is still present over eastern Europe but extends less to the north. The low-level front still extends from central Algeria to the Alpine region.

A broad positive PV anomaly over the United Kingdom and northwestern France

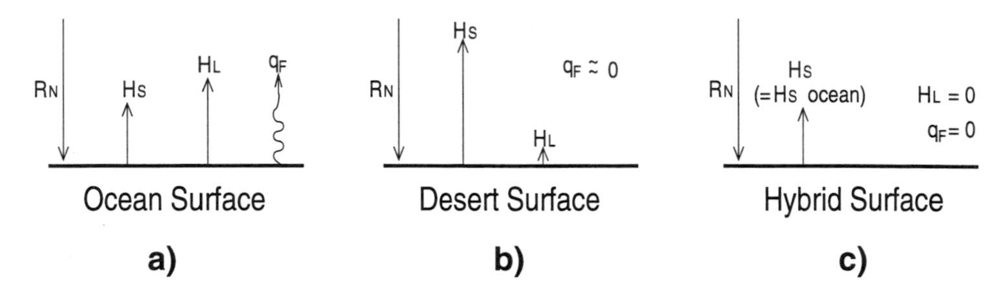

Figure 8.5: Schematic representation of the surface energy budget during daytime for (a) an ocean type surface, (b) a desert type surface and (c) the hybrid type surface used in the experiments. R_N, H_S and H_L represent the net incoming radiation, the sensible and latent heat fluxes, respectively. q_F represents water vapor evaporated into the atmosphere. The ground heat flux into the medium is omitted for clarity. See text for more details.

characterizes the upper-level synoptic situation at the beginning of the episode at 12 UTC 11 October (Fig. 8.4 a). This broad trough evolves into a narrow meridionally elongated PV streamer during the next 48 hours, reaching its maximum meridional extension deep inside Morocco at nearly 12 UTC 13 October (Fig. 8.4 c) before cutting-off from the main PV reservoir and settling over the western Mediterranean at 12 UTC 14 October (Fig. 8.4 d). The eastern flank of the anomaly reaches the western Mediterranean at some stage between 12 UTC 12 October (Fig. 8.4 b) and 12 UTC 13 October. Note that the flow over the Alps on the 320 K isentropic surface during this period changes from southwesterly (Fig. 8.4 a) to southeasterly (Fig. 8.4 d).

8.2 Methodology

Two elements require special care when suppressing moisture fluxes from a water basin: the characteristics of the surface used to replace the water-type surface and the spin-up time necessary to evacuate moisture present over the water basin.

8.2.1 Surface Type Change of the Water Basins

The modification of the surface type of the water basins must not modify the energy budget at the surface. At a surface of ocean type (Fig. 8.5 a), a balance is reached between the net incoming radiation (R_N) and the sensible and latent heat fluxes (H_S and H_L, respectively). The Bowen ratio (defined as $B = H_S/H_L$) is less than one in this case. An amount of moisture equivalent to the latent heat

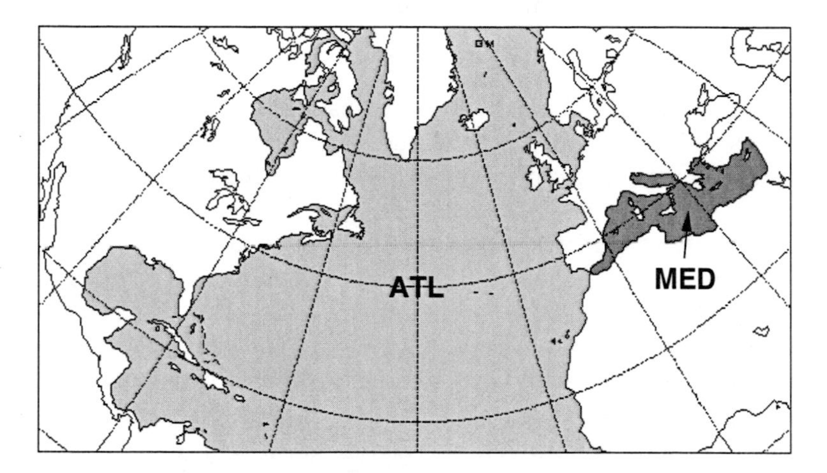

Figure 8.6: Extension of the two water basins that have been modified in the experiments of Part II. The Atlantic water basin (ATL) is depicted by the light grey area and the Mediterranean (MED) by the dark grey area.

flux H_L is also released into the atmosphere (q_f). A desert type surface (Fig. 8.5 b) will hardly supply any moisture into the atmosphere. However, this type of surface will modify the energy budget by increasing the sensible heat flux H_S instead of the latent heat flux H_L (i.e. the Bowen ratio will be larger than unity) that can in turn warm and destabilize the boundary layer[1]. In this study, the Climate High Resolution Model (CHRM, c.f. Section 2.3.1) is used in which H_S is a function of the temperature at the surface (T_0) and at the lowest model level while H_L is a function of the specific humidity of saturation at the surface at T_0 and of the specific humidity at the lowest model level. In order to eliminate the evaporation, a hybrid surface (Fig. 8.5 c) is utilized in which H_S takes on the same value as for the ocean surface (and subsequently the same T_0) and H_L is set to zero so that the evaporation rate q_f is also zero. Note that the modification of H_L does not modify H_S in this case.
The regions where the change of type of the ocean surface has been made are shown in Fig. 8.6. Note that for each set of experiments, four simulations are conducted: one control simulation (CTRL hereafter), one with a modified Atlantic surface (noATL), one with a modified Mediterranean surface (noMED) and one with modified surfaces of both the Atlantic and the Mediterranean (noATLMED).

[1]See also Arya (1998) for more details about surface energy budgets

8.2.2 Time-Spatial Considerations

A second difficulty arises when the large dimensions of the Atlantic are considered. The change of the surface type of the water basin does not immediately influence the moisture present directly above and produced by former evaporation from the same water basin. This remnant moisture has to be evacuated either by precipitation or by advection prior to any analysis, so that the flow arriving in the Alpine region does not hold moisture evaporated from the water basin prior to the modification of its surface type. In the case of the Atlantic, at least five model days are necessary in order to advect away the remnant moisture. This required 'spin-up' period in turn noticeably extends the duration of the simulation to the point that the main synoptic features may not be satisfactorily simulated towards the end of the simulation period. As the interest is set on the precipitation in the Alpine region, this difference in the flow pattern (relative to the analysis data) would make it difficult to distinguish the actual relevance of the moisture from the differences due to the discrepancies in the flow field induced by the longer simulation period. In order to address this issue, a two-step procedure is adopted (see Fig. 8.7).

In a first step and for both flood events, four series of six-hour simulations (CTRL and 3 experiments) are run on a large domain encompassing both the Northern Atlantic and the Mediterranean (Fig. 8.8) with a horizontal resolution of 1°. The simulations are started at 00 UTC 13 September 1993 for the September event and at 00 UTC 3 October 2000 for the October flood event. For each 6-hour simulation, the surface of the corresponding water basin is modified. The moisture fields (i.e. the specific humidity) of each +6h simulation are then inserted into the analysis data of the corresponding date that will serve as the initial conditions for the next 6-hour simulation (e.g. Fig. 8.7). The resulting initial conditions are likely to be imbalanced but the model output at +6h (which is relevant for the next step) will be balanced by the model. The result of this split procedure is a gradual 'drying' of the atmospheric layers above the considered water basin. This first step is conducted over 13 days (i.e. ending at 00 UTC 26 September 1993 and to 00 UTC 16 October 2000 for each flood event, respectively) for each experiment.

In a second step, outputs from the first step are used to initialize two types of simulations. One type are five sets of 72-hour numerical simulations run on a smaller domain (see Fig. 8.8) centered on the Alpine region with a higher horizontal resolution (0.5°). These simulations use the outputs of the first step as initial and boundary conditions for the simulations on the smaller domain. Figure 8.7 schematically shows the simulation build-up. To cover the major part of the flood events, the following sets of simulations have been conducted: two for the September 1993 episode, started at 00 UTC 22 September (BRIG1) and at 00 UTC 23 September (BRIG2) and three for the October 2000 episode, started

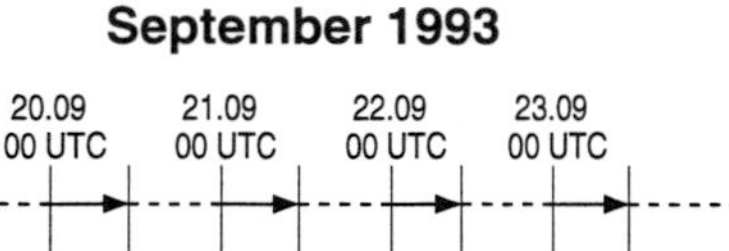

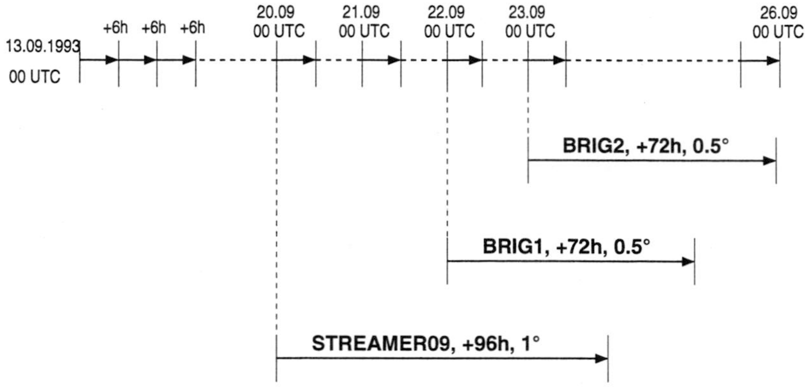

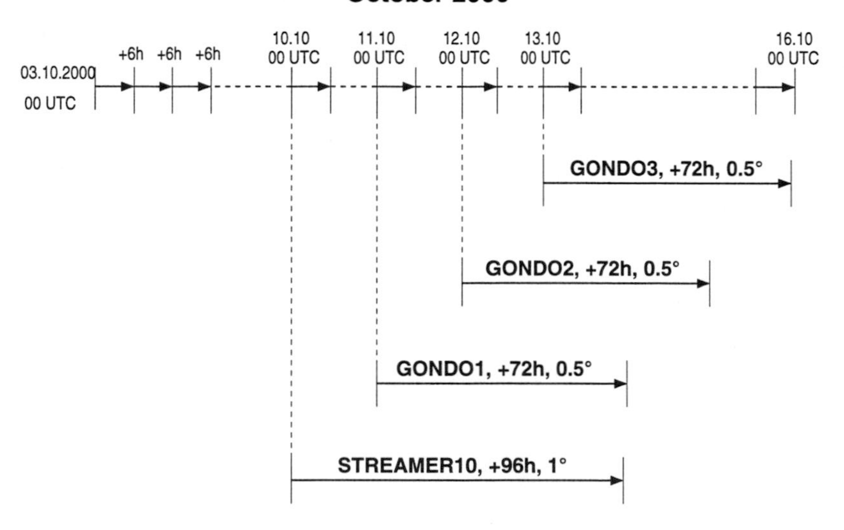

Figure 8.7: Schematic summary of the simulation procedure. The upper (lower) scheme applies for the September 1993 (October 2000) flood event. In each scheme, the upper series of small arrows represent the series of 6-hour simulations of the first step. The humidity fields produced by a +6h simulation are inserted into the initialization fields of the next +6h run. The longer arrows underneath indicate the simulations of the second step and their initial conditions (and starting time) correspond to those of the date of the first step linked by the vertical dashed line. On each of these long arrows, the name, duration and horizontal resolution of the set of simulations (CTRL plus 3 experiments) is indicated, 1° (0.5°) corresponding to the resolution of the large (small) domain in Fig. 8.8.

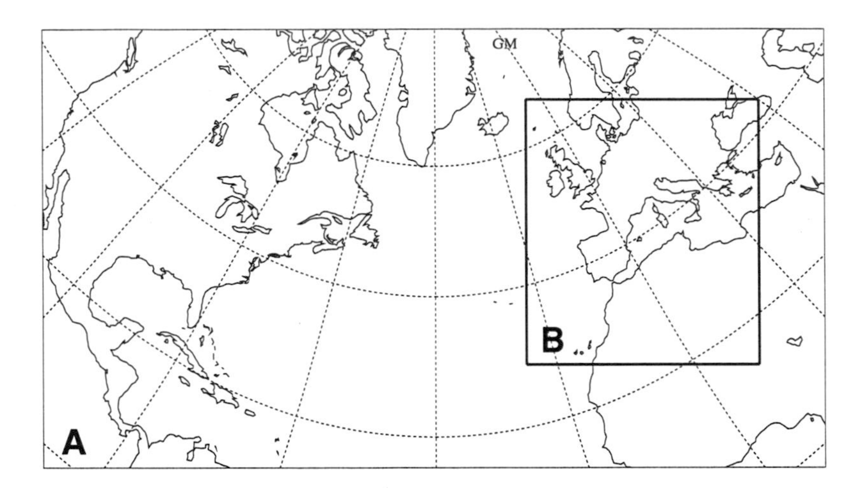

Figure 8.8: The two domains used in the numerical experiments. Domain A has a 1° horizontal resolution. Domain B has a 0.5° horizontal resolution.

at 00 UTC 11 October (GONDO1), at 00 UTC 12 October (GONDO2) and at 00 UTC at 13 October (GONDO3). The results of this group of simulations are discussed in Chapter 9. A second type of simulations uses the larger domain and the ouputs of the runs of the first step for the initial and boundary conditions. Here, one set of 96-hour simulations is started at 00 UTC 20 September (STREAMER09) and at 00 UTC 10 October (STREAMER10). Results are discussed in Chapter 10.
Note that evaporation from land surfaces can also be relevant, depending on the surface type and vegetation type (e.g. Heck 1999 and references therein). However, this aspect will not be considered in this study as the focus is on moisture transported from the Atlantic or the Mediterranean.

Chapter 9

Impacts on Precipitation

In this Chapter, the results of sensitivity studies are presented where the evaporative capability of the Atlantic and/or of the Mediterranean has been removed for the September 1993 and October 2000 heavy precipitation episodes. The variation of the intensity and distribution of the precipitation are analyzed in each episode. An alternative illustration of the respective roles played by the Atlantic and the Mediterranean for the amounts of Alpine rainfall is presented using the factor separation method of Stein and Alpert (1993). A search is also made for other moisture sources in the October 2000 episode. Finally, the relationship between the variability of the moisture source distribution and the structural characteristics of the co-developing upper-level PV streamers is explored.

9.1 The Brig Flood Event

The Brig episode of heavy precipitation lasted from 22 September to 25 September 1993. The focus will be set on the middle period (23-24 September) and only the situation for the 23rd September will be shown here, as it is representative of the whole episode. This corresponds to the +24h-+48h time interval of the BRIG1 set of experiments (e.g. Chapter 8). The same type of results have been calculated for the 24th September from the BRIG2 set of experiments and can be found in Fig. E.1 in the Appendix.

The control (CTRL) simulation shows maxima exceeding 80 mm in the 24-hour accumulated rainfall for the 23rd September 1993 on the Alpine southside south of Switzerland and in the French Maritims Alps (Fig. 9.1 CTRL). Other maxima are found over Sardinia and northeastern Algeria. These results agree in position and intensity with those of the ECMWF forecasts (e.g. Fig. D.1 b in the Appendix), except for the rainfall amounts over the Pyrenees and the Massif

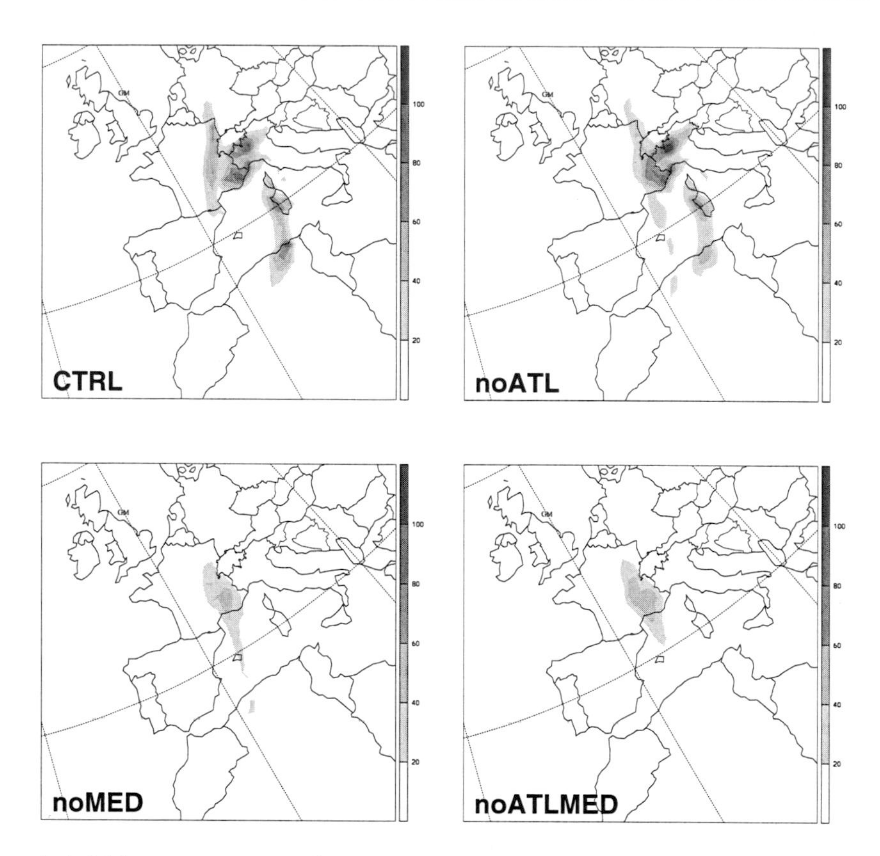

Figure 9.1: 24-hour accumulated precipitation in millimeters on 23 September obtained from the BRIG1 experiment. CTRL, noATL, noMED and noATLMED refer to the control simulation and to the simulations without evaporation from the Atlantic, the Mediterranean and without evaporation from both the Atlantic and the Mediterranean, respectively.

Central that are less in the model and the maximum south of the Swiss canton of Ticino that is shifted some 100 kilometers farther eastward.
The simulation without evaporation from the Atlantic (noATL, Fig. 9.1) is very similar to the CTRL simulation and exhibits a slight increase in the rainfall amounts in the Alpine region south of Switzerland and from the French Maritimes Alps to northern Algeria.
The absence of evaporation from the Mediterranean (Fig. 9.1 noMED) results in dramatic changes in the precipitation pattern compared to the CTRL simulation. In effect, there is a quasi-disappearance of the precipitation in the Alpine region south of Switzerland and in the Po Plain. No rainband is found from Corsica to northeastern Algeria anymore and there is an important decrease of the rainfall amounts also in southeastern France.
The removal of the evaporation from both the Atlantic and the Mediterranean

(noATLMED, Fig. 9.1) gives results similar to the noMED simulation with a quasi-disappearance of the precipitation on the southern Alpine slopes.
The main role of the Mediterranean and the most secondary role of the Atlantic as moisture sources for the Alpine rainfall amounts can also be identified with horizontal water vapor fluxes. Figure 9.2 shows the low-level horizontal moisture flux distribution vertically integrated between the surface and 800 hPa at 00 UTC on 23 September for the BRIG1 simulation set. The CTRL simulation is characterized by large values of southerly moisture fluxes over the western Mediterranean. The noATL simulation presents a similar pattern with weaker moisture fluxes over the Atlantic. The simulations without evaporation from the Mediterranean (noMED and noATLMED), in comparison to the CTRL simulation, exhibit a dramatic decrease of the southerly low-level moisture flux over the western Mediterranean, especially from the Gulf of Genoa to northern Tunisia. This situation is found throughout the episode, from 00 UTC 23 September to 00 UTC 24 September. The 24-hour evolution of the horizontal water vapor flux in this period can be found in Fig. E.3.
The situation for the 24th September is essentially similar to the situation of the 23rd September with the moisture of the Mediterranean having the main impacts on precipitation. 24-hour accumulated precipitation charts for the BRIG2 set of experiments analogous to Fig. 9.1, as well as horizontal water vapor fluxes charts analogous to Fig. 9.2 and E.3 can be found in Figs. E.1, E.2 and E.4, respectively. Note that these results contrast with those obtained by Romero et al. (1997) who investigated the roles of orography and evaporation from the Mediterranean in a heavy rainfall episode in Catalonia (Spain) in November 1988. They found the effect of the evaporation to be negligible for precipitation (their Figures 9 a and 13 b). However they concluded that evaporation might be an important factor, but for periods longer than their simulation (12 hours). The results of the present study also suggest the necessity of the 'pre-drying' process applied here in which the moisture present in the atmosphere produced by former evaporation is mostly removed.

9.2 The Gondo Flood Event

The Gondo event lasted longer than the Brig episode (from 11 October to 16 October). Here, the focus is set on the middle period (12-14 October 2000). This corresponds to the results of the GONDO1, GONDO2 and GONDO3 sets of experiments, respectively.

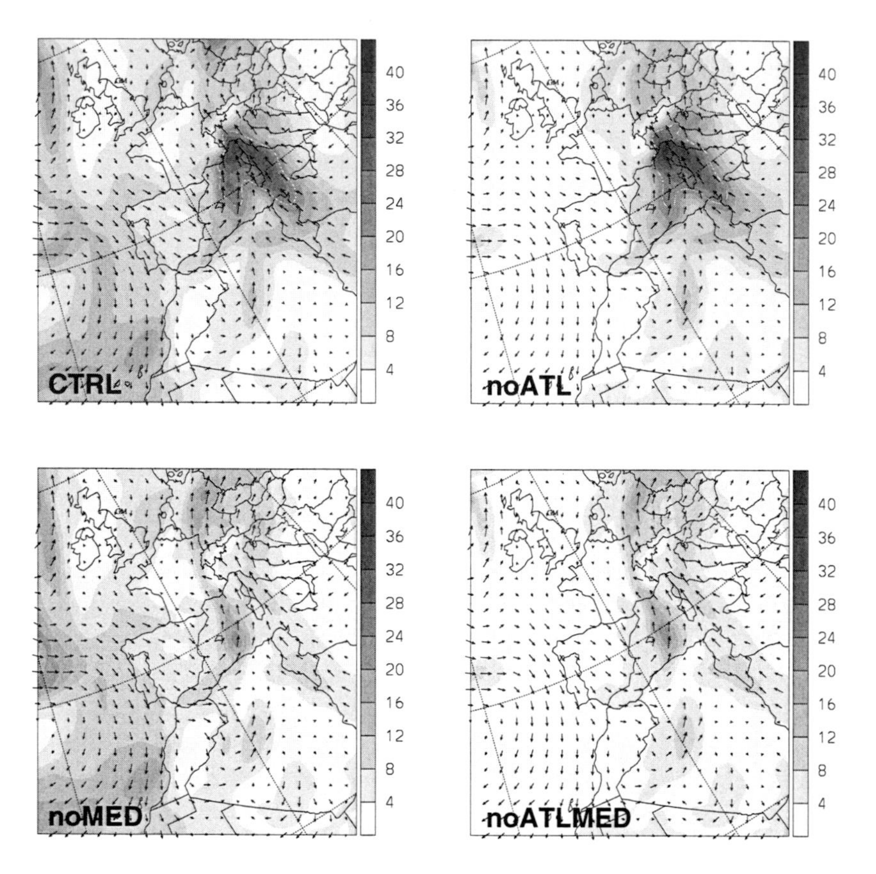

Figure 9.2: Horizontal moisture fluxes vertically integrated from 1000 hPa to 800 hPa (shaded, unit: $kg\ [H_2O]/s^3$) and wind vectors at 850 hPa. The white dashdotted line outlines the 24 $kg\ [H_2O]/s^3$ isoline. Results are for 12 UTC on 23 September, from the BRIG1 experiment.

9.2.1 12 October 2000

A maximum in the precipitation intensity is found in the region of Trieste in the CTRL simulation (Fig. 9.3 CTRL). It constitutes the northeastern part of a band of 24h-rainfall accumulation extending from the Balearic Islands. Rainfall amounts exceeding 20 mm are found north of Morocco and in the Bay of Biscay. Note that the rainfall amounts given by the model agree with those of the ECMWF forecasts (e.g. Fig. D.2 a in the Appendix), except for the maximum values over northeastern Italy that are larger. There is also a shift of some 300 kilometers to the east of the position of the maximum in comparison to the ECMWF forecasts.

The simulation without evaporation from the Atlantic exhibits a large decrease of up to 80 mm (Fig. 9.3 noATL) of the rainfall amounts in the region of Trieste

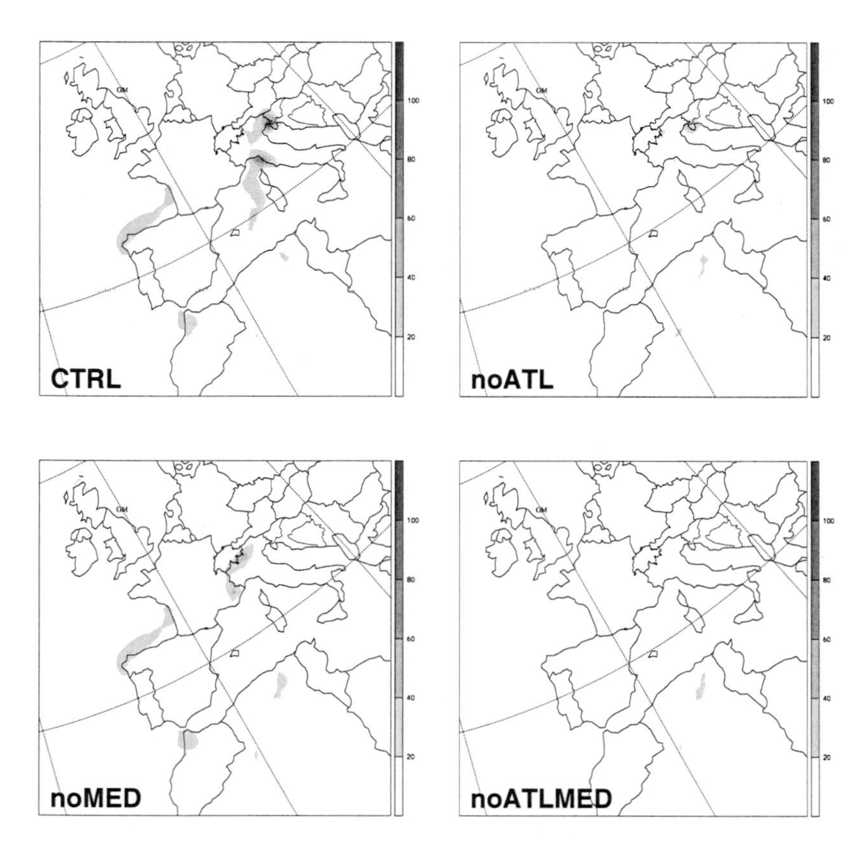

Figure 9.3: 24-hour accumulated precipitation in millimeters on 12 October obtained from the GONDO1 experiment. CTRL, noATL, noMED and noATLMED refer to the control simulation and to the simulations without evaporation from the Atlantic, the Mediterranean and from both the Atlantic and the Mediterranean, respectively.

as well as a general decrease of the precipitation intensity in the Po Plain. The rain band extending from the Ligurian Sea to the Balearic Islands present in the CTRL simulation has totally vanished, as well as the rain zones over northern Morocco and the Bay of Biscay.
The maximum in the rainfall amounts visible in the CTRL simulation over the region of Trieste has disappeared in the simulation without evaporation from the Mediterranean (Fig. 9.3 noMED). Nevertheless the rain amounts have slightly increased in the Alpine region, especially in the region of the French Maritimes Alps. The band of rain extending from the Ligurian Sea to the Balearic Islands has totally disappeared, as previously noticed in the noATL simulation. This suggests the role played by moisture from the Atlantic to favour the saturation of the air over the Mediterranean and subsequent condensation and precipitation. Therefore, the Atlantic and the Mediterranean are here both relevant moisture sources for the precipitation.

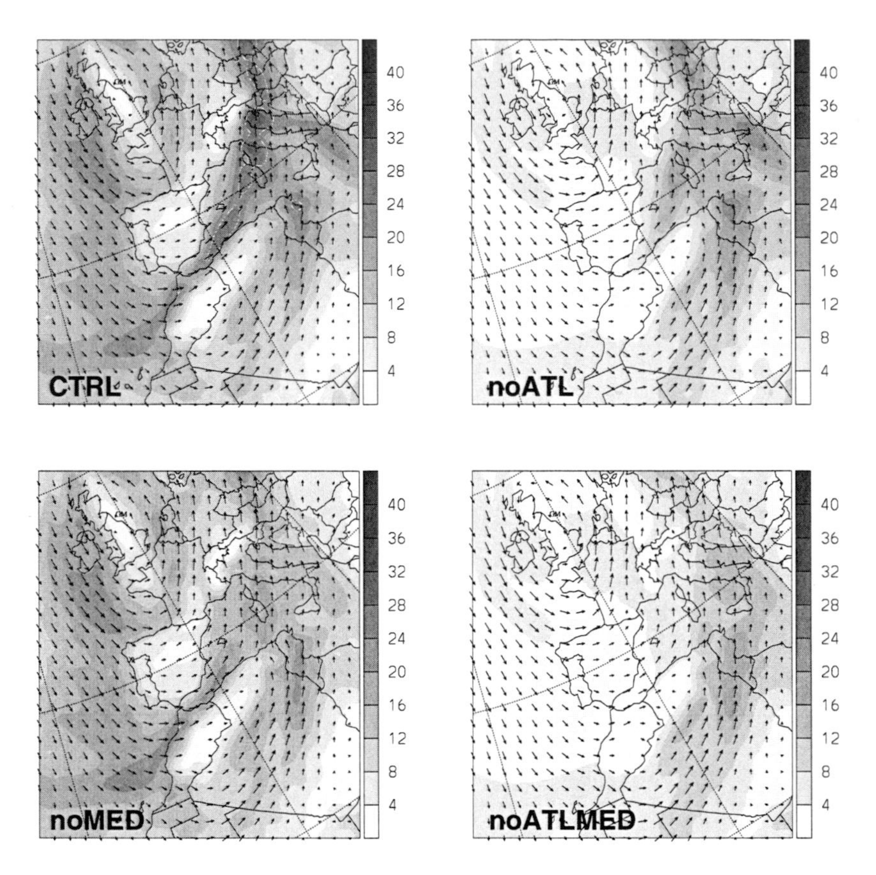

Figure 9.4: Horizontal moisture fluxes vertically integrated from 1000 hPa to 800 hPa (shaded, unit: $kg\ [H_2O]/s^3$) and wind vectors at 850 hPa. The white dashdotted line outlines the 24 $kg\ [H_2O]/s^3$ isoline. Results are for 12 UTC 12 October, from the GONDO1 experiment.

The simulation without evaporation from both the Atlantic and the Mediterranean (Fig. 9.3 noATLMED) is characterized by the quasi-absence of rainfall amounts in the Alpine region and over the western Mediterranean

The differences in the rainfall amounts for 12 October 2000 are related to the differences in southerly moisture fluxes toward the Alps. Figure 9.4 shows the low-level horizontal moisture flux distribution vertically integrated between the surface and 800 hPa at 12 UTC 12 October 2000 for the four experiments. The CTRL simulation is characterized by large values of moisture fluxes in a band extending from Morocco to Central Europe. Wind vectors at 850 hPa are southwesterly. In the noATL and noMED simulations, moisture fluxes are clearly weaker over the western Mediterranean south of the Alps. Finally in the noATLMED simulation, the southerly moisture flux toward the Alps is nearly negligeable (Figure E.5 in the Appendix shows the 24-hour evolution of the moisture flux

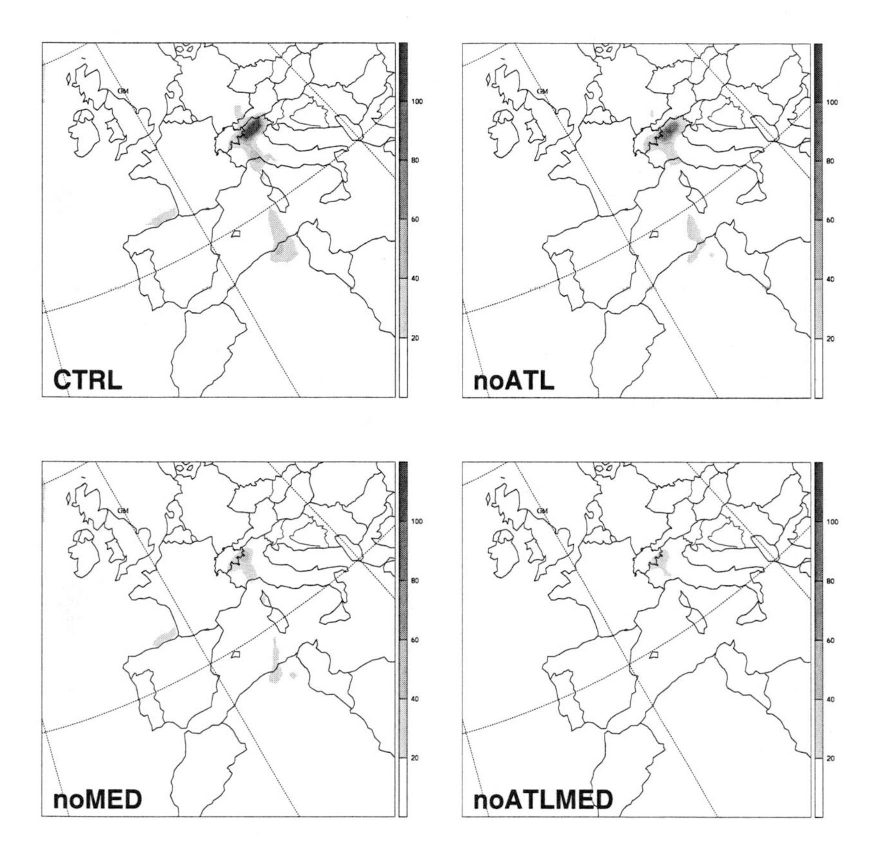

Figure 9.5: Same as Fig. 9.3 but on 13 October from the GONDO2 experiment.

distribution between 12 October 00 UTC and 13 October 00 UTC).

9.2.2 13 October 2000

The main feature in the daily accumulated precipitation of the CTRL simulation (Fig. 9.5 CTRL) is the extreme values of rainfall amounts found over the southern Julian Alps. Patterns of large values form a rainband from the Po Plain to northeastern Algeria. Some precipitation are also found along the north coast of Spain. The magnitude and position of the precipitation calculated by the model are in good agreement with the precipitation forecasted by ECMWF (e.g. Fig. D.2 b), except for the maximum that is again shifted some 300 kilometers farther to the east and larger.
Differences between the CTRL and noATL simulations are not very relevant (Fig. 9.5 noATL). There is a slight decrease of the rainfall amounts in the southern Alpine region in the ATL simulation and in the rainband over the western

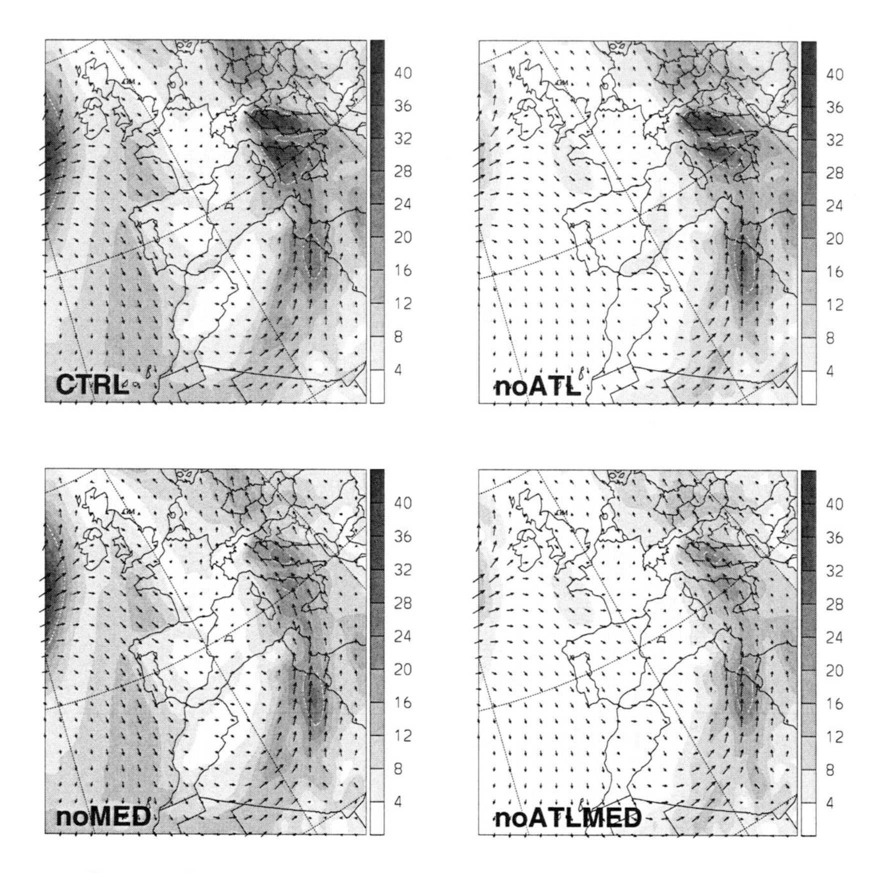

Figure 9.6: Same as Fig. 9.4 but at 12 UTC 13 October from the GONDO2 experiment.

Mediterranean. The rainband along the north coast of Spain has disappeared.
The simulation without evaporation from the Mediterranean (Fig. 9.5 noMED) exhibits a large decrease of the rainfall amounts over the southern Alpine slopes and the local maximum present over the southern Julian Alps in the CTRL and noATL simulations has disappeared. A slight decrease in the accumulation of precipitation is also visible over northeastern Algeria.
The combined removal of the evaporation from the Atlantic and from the Mediterranean (Fig. 9.5 noATLMED) induces a strong decrease of the rainfall amounts in the southern Alpine region, comparable in magnitude to the one found in the noMED simulation. Moreover, the band of precipitation extending from the Ligurian Sea to northeastern Algeria has disappeared, like the rain zone along the north coast of Spain.

The low-level horizontal moisture fluxes are consistent with the differences found in the rainfall amounts (Fig. 9.6). The CTRL simulation at 12 UTC exhibits large southerly moisture fluxes over Italy and south of Tunisia. On the contrary to 12 October, the offset of the evaporation from the Atlantic (noATL) does not

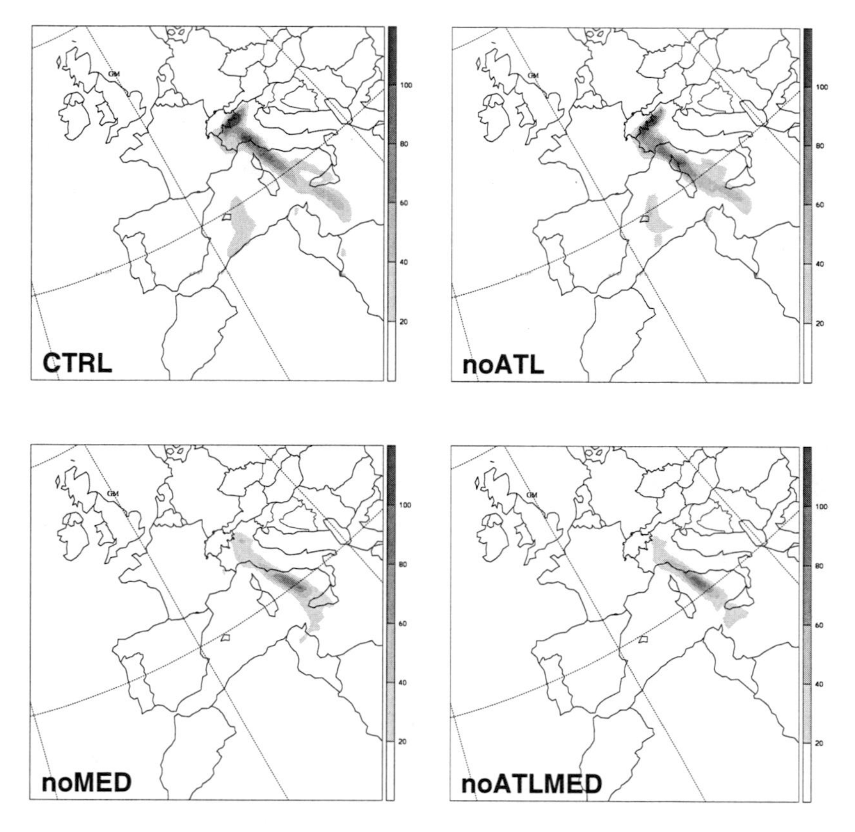

Figure 9.7: Same as Fig. 9.3 but on 14 October from the GONDO3 experiment.

modify the moisture fluxes in the lower levels in a relevant manner. However, the two simulations without evaporation from the Mediterranean (noMED and noATLMED) show weaker moisture fluxes over Italy. The moisture flux over Tunisia does not exhibit particular differences between the two simulations.
The time evolution of the moisture flux between 00 UTC 13 October and 00 UTC 14 October shows an interesting feature (Fig. E.6). At 00 UTC 13 October, the differences in the moisture flux over the western Mediterranean between the CTRL and noATL are still visible, while they can hardly be identified at 12 UTC and at 00 UTC 14 October. This suggests that the direct impact of the Atlantic moisture on the rainfall amounts in the southern Alpine region decreases rapidly between 00 UTC and 12 UTC on 13 October.

9.2.3 14 October 2000

The CTRL simulation (Fig. 9.7 CTRL) shows very large amounts of precipitation (over 100 mm) over southeastern Switzerland. Large amounts of rainfall are also found in the Po Plain, the Tyrrhenian Sea and along the west coast of Libya. Rain zones are also visible south and east of the Balearic Islands and west of Spain. Compared to the ECMWF forecasts (Fig. D.2 c), the maximum is shifted farther eastward (some 200 kilometers) and overestimated (40 mm more). It is noteworthy that the pattern of the rainbands also differs quite significantly. The ECMWF forecast displays a two-band structure with the first one extending from northeastern Algeria to western Switzerland and the second shorter one from the eastern coasts of Tunisia to Sardinia. In the CTRL simulation, a first band of rainfall extends from east of Tunisia to the Alpine region and a second zone is visible east and south of the Balearic Islands.

The intensity of the precipitation in the case where no evaporation from the Atlantic is available (Fig. 9.7 noATL) does not differ from the CTRL simulation, except for the disappearance of the weak rain zone west of Spain.

The noMED simulation (Fig. 9.7 noMED) exhibits a significant decrease of the rainfall amounts over the southern Alps. A large decrease in the amount of precipitation is also visible in the band extending from the Po Plain to east of Tunisia and the rain zone in the vicinity of the Balearic Islands has totally disappeared. Note however the presence of large amounts of precipitation in the Tyrrhenian Sea.

The simulation without evaporation from both the Atlantic and the Mediterranean (Fig. 9.7 noATLMED) has a rain pattern similar to the one of the noMED simulation and the maximum in the Tyrrhenian Sea is still present.

The moisture flux on the Alpine southern slopes is mainly southeasterly at 12 UTC on 14 October with maximum intensity over the Tyrrhenian Sea in the CTRL simulation (Fig. 9.8). The intensity and direction of the moisture flux pattern south of the Alps in the noATL simulation are quite similar to those of the control simulation. The removal of the evaporative capability of the Mediterranean (noMED and noATLMED) significantly diminishes the intensity of the moisture flux over the Tyrrhenian Sea. The time evolution of the moisture flux pattern on 14 October (Fig. E.7) shows a larger impact of the removal of the evaporation from the Mediterranean compared to the Atlantic. Large values of the moisture flux are present from Italy to the African continent in the noATLMED during the whole day.

Note that some large amounts of precipitation and strong moisture fluxes are still visible from the Po Plain to Sicily, despite the removal of the evaporative abilities of the Atlantic and the Mediterranean (Fig. 9.7 and E.7 noATLMED). This suggests the possible implication of a third moisture source. This aspect will be explored in Section 9.4

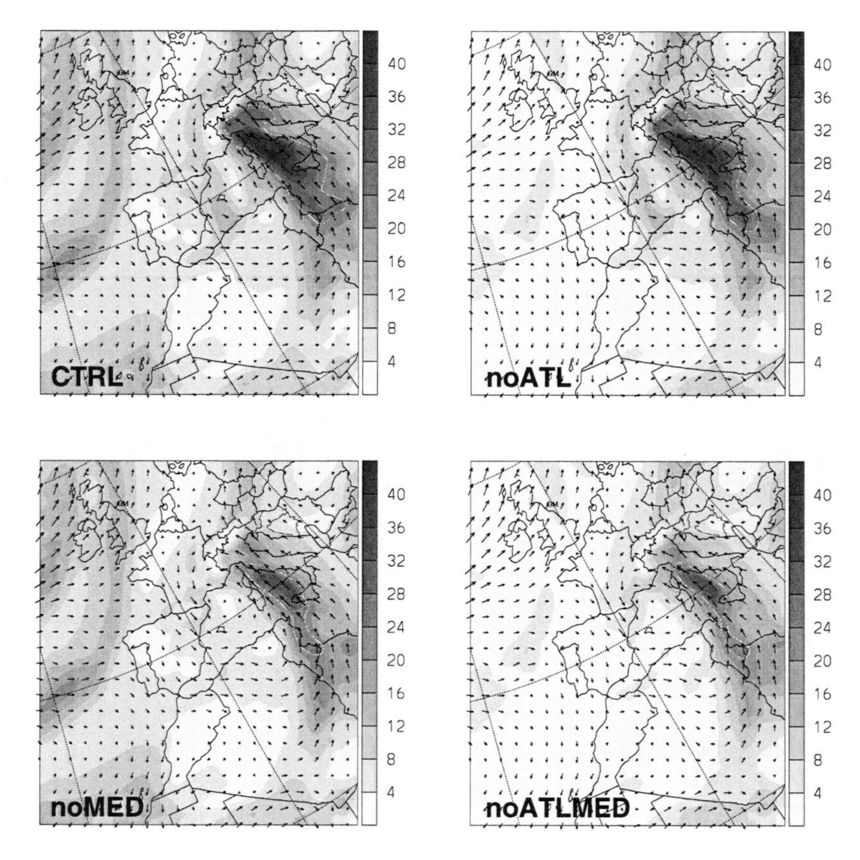

Figure 9.8: Same as Fig. 9.4 but at 12 UTC 14 October from the GONDO3 experiment.

9.3 Relative Contributions

In Sections 9.1 and 9.2 the effects of the removal of evaporation from the Atlantic and the Mediterranean have been studied. In the present Section, an alternative illustration is sought in which precipitation patterns are directly attributed to each water basin. However, their interaction is nonlinear, making it difficult to assess directly which part of the total precipitation can be attributed to which water basin. The factor separation method proposed by Stein and Alpert (1993) is used hereafter in an attempt to isolate the effects of moisture coming from the Atlantic or from the Mediterranean. This technique was used in several studies to isolate the individual contributions and the interactions of several factors (Balasubramanian and Yau 1995, Alpert et al. 1996, Morgenstern and Davies 1999, Tsidulko and Alpert 2001). According to this method, 2^n experiments are needed in order to study the effects of n factors.
In our case, four experiments are undertaken to examine the effects of the Atlantic

and of the Mediterranean. If $\mathcal{F}_c$ denotes the CTRL simulation (with the evaporation of both the Atlantic and the Mediterranean available), $\mathcal{F}_{na}$ the noATL simulation (evaporation from the Atlantic switched off), $\mathcal{F}_{nm}$ the noMED simulation (evaporation from the Mediterranean switched off) and $\mathcal{F}_{nam}$ the noATLMED (evaporation from both water basins suppressed), then the factor separation method yields:

(i) relative precipitation attributable to the Atlantic moisture: $\mathcal{G}_a = \mathcal{F}_{nm} - \mathcal{F}_{nam}$,

(ii) relative precipitation attributable to the Mediterranean moisture: $\mathcal{G}_m = \mathcal{F}_{na} - \mathcal{F}_{nam}$,

(iii) relative precipitation attributable to the interaction between the moisture of both origins: $\mathcal{G}_{am} = \mathcal{F}_c - \mathcal{F}_{na} - \mathcal{F}_{nm} + \mathcal{F}_{nam}$,

(iv) relative precipitation unrelated to the moisture of both origins: $\mathcal{G}_0 = \mathcal{F}_{nam}$

The term $\mathcal{G}_0$ corresponds exactly to the noATLMED simulations and will not be discussed in the present Section. The physical interpretation of the terms $\mathcal{G}_a$, $\mathcal{G}_m$ and $\mathcal{G}_0$ are intuitively relatively clear. However, the unubiquitousness of the interpretation of the term $\mathcal{G}_{am}$ is more complicated to establish. One can think, for instance, of the effect of Atlantic air arriving over the Mediterranean. Depending on the water vapor content of this air, evaporation from the Mediterranean and/or the conditions influencing saturation and subsequent precipitation may vary.
Sections 9.3.1 and 9.3.2 indicate the 'pure' effects of the Atlantic and Mediterranean moisture and the effect of their interaction on 24-hour accumulated precipitation intensity and distribution for the Brig and Gondo episodes.

9.3.1 The Brig Episode

The results of the factor separation method for the 23rd September are shown in Fig. 9.9, together with the precipitation obtained by the CTRL simulation. It appears that there is no effect of the Atlantic moisture on the precipitation (Fig. 9.9 b). Conversely, most of the rainfall in the Alpine region and from Algeria to Corsica can be attributed to the Mediterranean (Fig. 9.9 c). The interaction between the Atlantic and the Mediterranean does not influence in a relevant manner the amounts or position of the rainfall in the Alpine region (Fig. 9.9 d) and seems to be mainly responsible for the rainband extending over France from the Pyrenees to the Alsace. The results for the 24th September (Fig. E.8 in the Appendix) give the same passive role of the Atlantic and the main role of the Mediterranean for the precipitation in the Alpine region. The interaction of both factors results

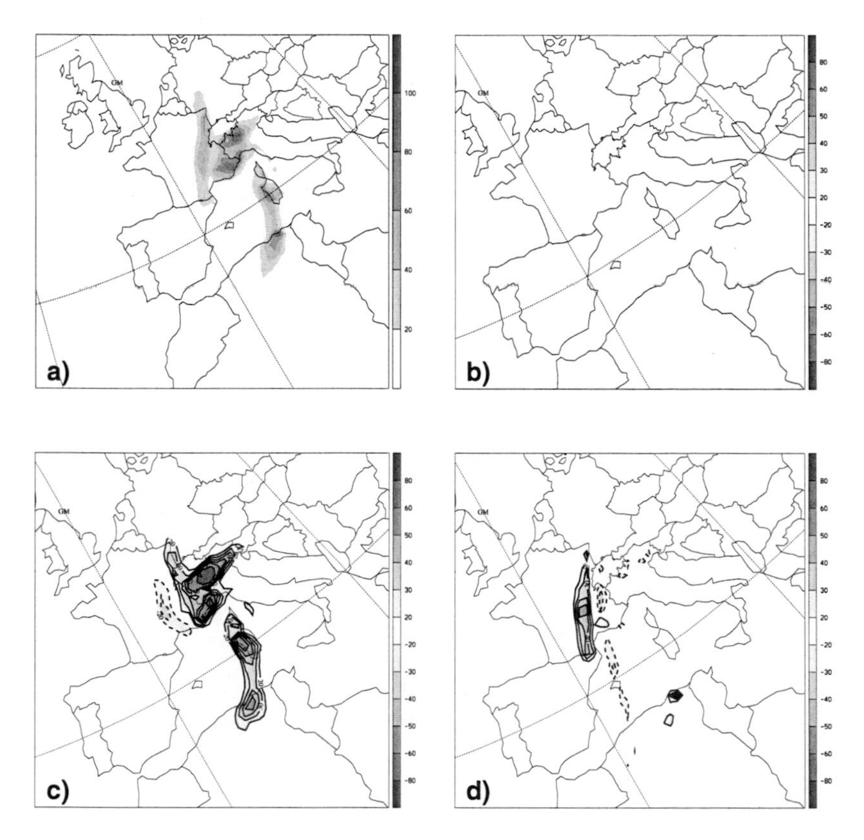

Figure 9.9: 24-hour accumulated precipitation in millimeters on 23 September attributable to (b) the Atlantic moisture ($\mathcal{G}_a$), (c) the Mediterranean moisture ($\mathcal{G}_m$), and (d) the interaction between the moisture from the Atlantic and from the Mediterranean $\mathcal{G}_{am}$. (a) shows the 24-hour accumulated precipitation in millimeters on 23 September as given by the CTRL simulation of the BRIG1 experiment.

in some localized enhancement of the rainfall amounts in the Tyrrhenian Sea and over northeastern Italy.

9.3.2 The Gondo Episode

Here, the results of the factor separation method are only presented for 12 October and 14 October. Results for 13 October can be found in Fig. E.9 in the Appendix.

On 12 October, the effect of the Atlantic moisture is to increase significantly the precipitation on the Alpine southside (in southern Switzerland and the Maritimes Alps, Fig. 9.10 b) and along the northern coasts of Spain. The 'pure' effect of the Mediterranean is to increase precipitation at the northeastern end of the Adriatic

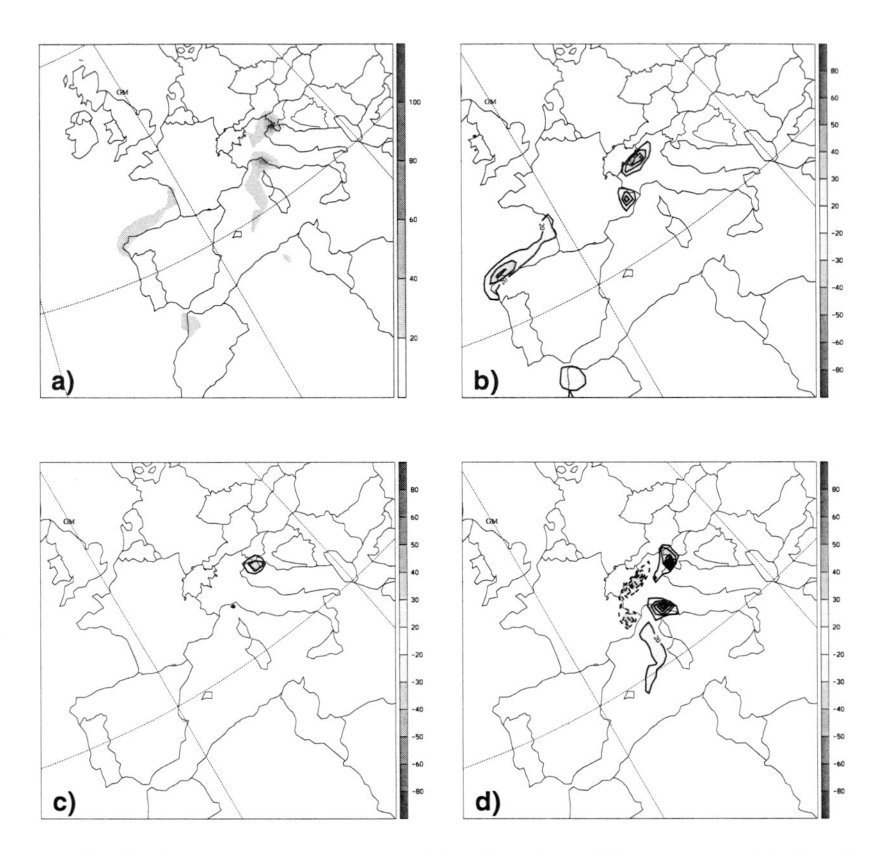

Figure 9.10: 24-hour accumulated precipitation in millimeters on 12 October attributable to (b) the Atlantic moisture ($\mathcal{G}_a$), (c) the Mediterranean moisture ($\mathcal{G}_m$), and (d) the interaction between the moisture from the Atlantic and from the Mediterranean $\mathcal{G}_{am}$. (a) shows the 24-hour accumulated precipitation in millimeters on 12 October as given by the CTRL simulation of the GONDO1 experiment.

Sea (Fig. 9.10 c). Note that no precipitation pattern in the Alpine region can be attributed to the Mediterranean moisture alone. The interaction between the Atlantic and Mediterranean moisture is here quite relevant (Fig. 9.10 d) as it is linked to a substantial decrease of the precipitation in the southern Alpine region (in southern Switzerland and the Maritimes Alps) and an increase of the rainfall amounts northeast of Corsica and in northeastern Italy.

The results are quite different for 14 October. The effects purely attributable to the Atlantic moisture are a slight increase of the precipitation in the Po Plain, as well as an eastward shift of the rainband over the Tyrrhenian Sea (Fig. 9.11 b). The effects due to the Mediterranean are comparable to the Brig case, in the sense that most of the precipitation falling on the southern side of the Alps can be attributed to the Mediterranean moisture (Fig. 9.11 c), as well as the precipitation east of the Balearic Islands. The Mediterranean moisture also plays

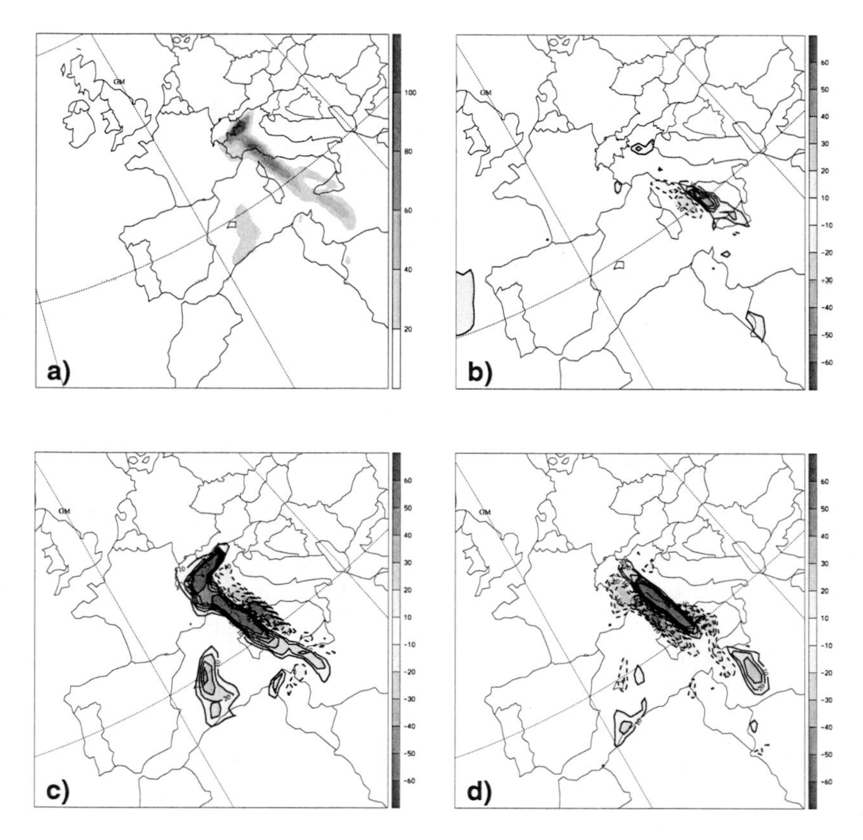

Figure 9.11: 24-hour accumulated precipitation in millimeters on 14 October attributable to (b) the Atlantic moisture ($\mathcal{G}_a$), (c) the Mediterranean moisture ($\mathcal{G}_m$), and (d) the interaction between the moisture from the Atlantic and from the Mediterranean $\mathcal{G}_{am}$. (a) shows the 24-hour accumulated precipitation in millimeters on 14 October as given by the CTRL simulation of the GONDO3 experiment.

a role in the position of the rainband extending from the Ligurian Sea to Sardinia as a westward shift is visible. Finally, the interaction between the Atlantic and the Mediterranean is visible through a shift to the east of the rainband extending from the Alps to Sardinia (Fig. 9.11 d) and an increase of the precipitation south of Sicily. Note that the results for 13 October (Fig. E.9) shows no 'pure' effect of the Atlantic on the precipitation (b). The Mediterranean contributes to the precipitation in the Alpine region in the Dolomites and the Friuli (c). The interaction between the Atlantic and the Mediterranean (d) results in an increase of the precipitation in the Friuli.
To summarize, the results of the factor separation for the Gondo event show a more complicated distribution of the effects attributed to the Atlantic and the Mediterranean than those found for the Brig event. 12 October is characterized by the influence of both the Atlantic and the Mediterranean on the precipitation

(in the individual effect as well as in the interaction effect). On 13 October, no effect of the Atlantic on the precipitation is visible. As time progresses, the relevance of the Mediterranean increases. On 14 October, the role of the Atlantic is only visible in shifts in the positions of the rainbands. The heavy precipitation in the Alpine region are mostly fed by the Mediterranean. Thus, there is a flow transition on 13 October from a southwesterly flow regime where both the Atlantic and the Mediterranean are relevant for the precipitation at the beginning of the episode (12 October) to a southerly-to-southeasterly flow regime where the main moisture source for the heavy precipitation in the Alpine region is the Mediterranean (14 October).

9.4 Discussion

Results from Sections 9.1 and 9.2 show the complexity and the high variability of the distribution of the moisture sources involved in heavy precipitation episodes along the southern side of the Alps. An insight into this complexity was suggested by Krichak and Alpert (1998) in their study of the relevance of tropical moisture from the Arabian Sea for the chain of events that led to the November 1994 series of floods (Buzzi et al. 1998).
In the Brig flood event, most of the rainfall has been attributed to the Mediterranean. Conversely, during the Gondo flood event (Section 9.2) different moisture sources get involved. From the beginning of the episode to 13 October, the Atlantic and the Mediterranean have similar impacts on the precipitation in the Alpine region. The interest in the contribution of Atlantic moisture to heavy precipitation episodes in the Alpine southside has been recently brought to center stage by several studies. Reale et al. (2001) investigated the relevance of Atlantic tropical moisture carried by ex-hurricanes that recurved over the eastern Atlantic and underwent extratropical transition prior to their arriving in the vicinity of Europe. In Chapter 10 of Pinto (2002), a classification is made of thirty extreme precipitation events in Northern Italy based on the type on tropical-extratropical interaction over the North Atlantic. Either large amounts of moisture are carried directly across the Atlantic by the ex-tropical systems that underwent extratropical transition or the zonal wave pattern over the Atlantic is reinforced by the tropical-extratropical interaction and the subsequent formation of a trough over the eastern Atlantic favours southwesterly flow over the western Mediterranean. Other examples of the intrusion of Atlantic moisture in the western Mediterranean were also found during MAP IOP2B (Arnal 2003, Chapter 3).
The second half of the Gondo episode (from 13 October) is characterized by the rapid decrease of the impact of the Atlantic upon the rainfall amounts in the Alpine region. Nevertheless, some large amounts of precipitation are still present from the Po Plain to Sicily despite the removal of evaporation from the Atlantic

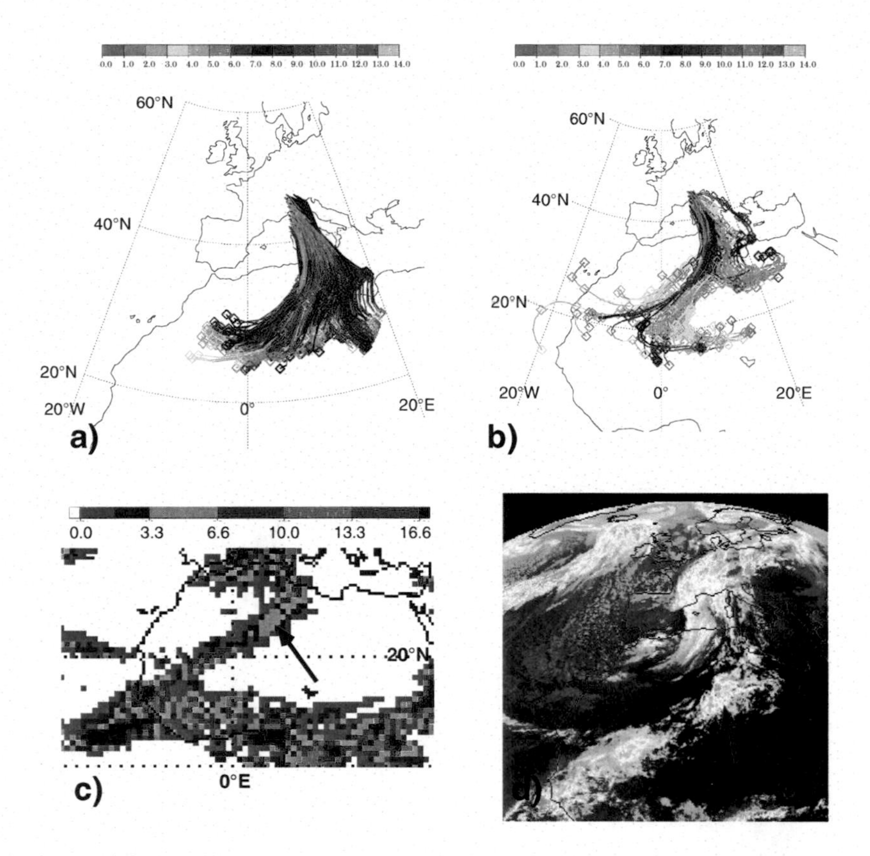

Figure 9.12: (a) 48-hour backward trajectories started between 950 hPa and 500 hPa from the domain encompassed within 8°E and 14°E and 40°N and 47°N. Calculations started at 00 UTC 15 October 2000 for points having a relative humidity larger than 80% and realized with the outputs of the noATLMED from the GONDO3 set of experiments. Colors indicate the specific humidity (in g/kg) along the trajectory. (b) 120-hour backward trajectories calculated from ECMWF analysis data. Initial domain, criterion and start time are the same as (a). (c) Combined daily accumulated rainfall in millimeters for 13 October 2000 from the Tropical Rainfall Measuring Mission (source: http://trmm.gsfc.nasa.gov). The black arrow indicates the rainfall region discussed in the text. (d) Meteosat infrared satellite imagery at 09 UTC 13 October. The red arrow shows the cloud region discussed in the text.

and the Mediterranean (e.g. Fig. 9.7 noATLMED). The 48-hour backward trajectories calculated from this CHRM simulation from the rainfall region (initial criterion: relative humidity larger than 80%) and started at 00 UTC 15 October suggest a North African origin of the moisture (Fig. 9.12 a). 120-hour backward trajectories calculated from ECMWF analysis data and started at the same date from the same region and with the same initial criterion (Fig. 9.12 b), point to the subtropical regions of Africa as a possible origin of the moisture. The presence of moisture over the Sahara between 13 October and 14 October is also visible in rain signals depicted on TRMM (Tropical Rainfall Measuring Mission) pictures (Fig. 9.12 c) and by the presence of cloud cover in satellite pictures (Fig. 9.12 d). The possible incursion of tropical moisture into the subtropics over Northern Africa has already been investigated by Knippertz et al. (2003) and Knippertz (2003). Note that a similar transport of moisture over the Sahara from the subtropics to the Mediterranean has also been identified in the MAP IOP2B (19-20 September 1999, e.g. Arnal 2003, Figure 3.12). However, the inspection of specific humidity along the back trajectories calculated from the noATLMED simulation output shows larger values over Algeria in comparison to the back trajectories calculated from the ECMWF analysis data. This is due to the fact that in the noATLMED (as well as in the three other simulations of the GONDO3 set) the model produces large amount of moisture in the mid- and lower troposphere over the Sahara. The cause of this unusual behaviour is still unknown and further studies with the calculation of horizontal moisture flux budgets in this region are needed in order to figure out the reason of this difference. Nevertheless, as already mentioned earlier, rainfall was detected over the Sahara that was not reproduced in the ECMWF analysis data.

Link Between the Upper-Level Precursor and Moisture Source Involvement

The variability in the impacts of the Atlantic and of the Mediterranean upon the amounts of rainfall between the Brig and Gondo heavy precipitation episodes can be linked to the evolution of the accompanying upper-level PV streamer.
In the Brig case, the meridionally elongated shape of the stratospheric intrusion is already formed as it reaches the western coasts of France at 00 UTC on 21 September. The PV streamer is then *advected* from the eastern Atlantic toward the Alpine region (e.g. Fig. E.10 for a four-day evolution of the PV streamer and lower-level baroclinic feature). As a consequence, its eastern flank and the associated front structure have already a nearly meridional orientation by the time they reach the Alpine region (Fig. 9.13 a). Note that the frontal structure remains oriented north-to south or northwest-to-southeast throughout the episode. By the time of the maximum meridional extension of the streamer, shortly before it cuts off from the main PV reservoir (06 UTC on 23 September, Fig. 9.13 b), its southern edge reaches the northern coasts of Morocco and Algeria. The baroclinic zone is extending from the Atlas Mountains to the Alpine region.
On the contrary, the streamer of the Gondo episode actually develops over western Europe (c.f. the four-day evolution in Fig. E.11). The orientation of the eastern flank of the streamer evolves from southwest-to-northeast between 11 October to nearly 00 UTC on the 13 October (e.g. Fig. 9.13 c) to south-to-north and

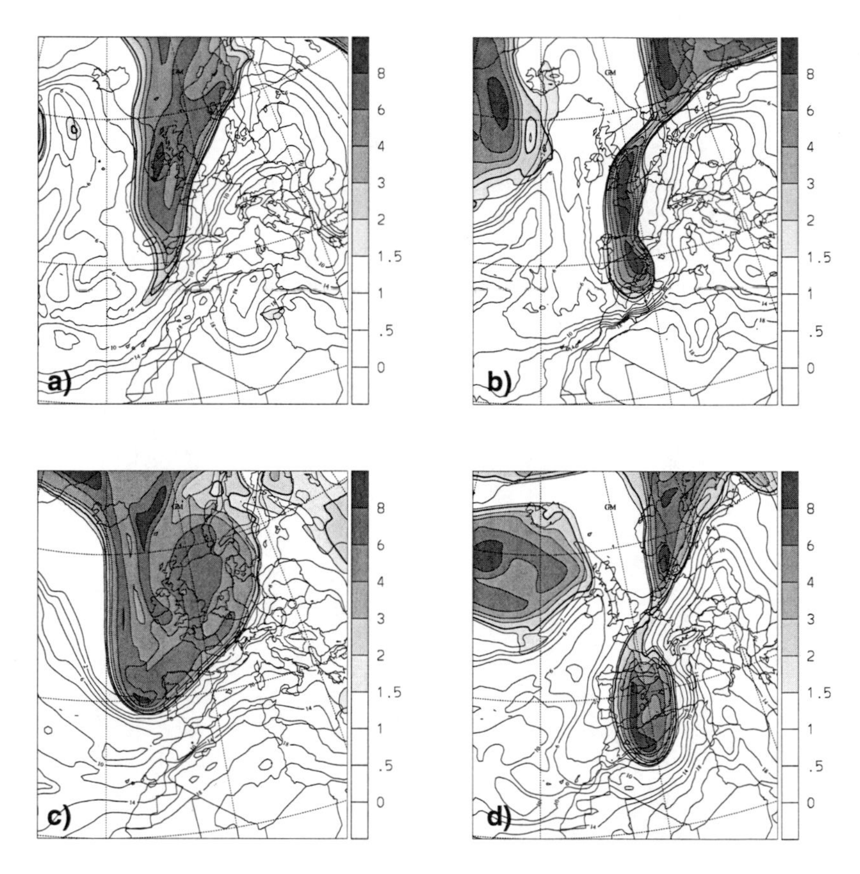

Figure 9.13: Distribution of potential vorticity in pvu (grey shaded) on the 320-K isentropic surface and temperature at 800 hPa derived from the ECMWF analysis data (solid lines, contours every 2 °*C* from 0 °*C* to 20° *C*) at (a) 00 UTC on 22 September 1993, (b) 06 UTC on 23 September 1993, (c) 00 UTC on 12 October 2000 and (d) 00 UTC 14 October 2000.

southeast-to-northwest during the second half of the episode. The orientation of the baroclinic zone at the eastern flank of the streamer follows a similar evolution. Moreover, the southern edge of the PV streamer reaches central Algeria by the time of its maximum meridional extension (14 October 00 UTC, Fig. 9.13 d), i.e. some 600 km farther south compared to the Brig case. The associated baroclinic zone extends from the Malian border to the Alpine region. It is suggested here that this farther southward extension of the frontal zone (and of the associated jet structure) sets the conditions for moisture of tropical or subtropical origins to be captured and re-directed northward toward the Mediterranean.
Hence, the variability in the evolution, structure and position of the PV streamer associated with heavy precipitation episodes on the Alpine southside (Massacand et al. 1998) can be linked to the variability in the distribution of the moisture

sources in extreme rain events in the Alpine region. The sensitivity of Alpine rainstorms to the structure of the meridionally elongated filament of intruded stratospheric air has already been suggested by Fehlmann et al. (2000) and the particular relevance of the southern part of the PV streamer for the prediction of such events was shown by Fehlmann and Quadri (2000). The results of this Chapter suggest that the determination of the moisture sources involved in heavy precipitation episodes is linked to the type of development (advection or development over western Europe) and to the meridional extension of the PV streamer. However, the genesis of the PV streamer (actually a Rossby wave breaking event) is a highly non-linear feature and many further studies will be necessary to try and figure out the many involved factors (see also Chapter 10).

9.5 Summary

In this Chapter, the impacts of the Atlantic and of the Mediterranean as moisture sources upon the precipitation of the Brig and Gondo flood events have been studied. A two-step method (described in Chapter 8) in which the atmosphere above the water basins is first 'dried' has been used. Although both heavy precipitation episodes share many common characteristics (southerly moist flow impinging upon the Alpine southside, presence of an upper-level meridionally elongated stratospheric intrusion over France), the roles played by both water basins differ, as well as the sequence in which they contribute to the rainfalls.
In the Brig event, the removal of the evaporation from the Mediterranean suppresses nearly all precipitation in the southern Alpine region. The factor separation method also attributes the moisture sources of the strong rainfalls mainly to the Mediterranean.
The Gondo event appears to have a more complex distribution of the moisture sources involved. In the first part of the episode, both the Atlantic and the Mediterranean contribute to the large amount of rainfall. On the 12th October precipitation in the Alpine region can be attributed to both water basins. From the 13th October, there is a gradual swing of relevance from the Atlantic to the Mediterranean to the extent that the role played by the Atlantic only amounts to an eastward shift of the rainband extending from the Alps to Sicily on the 14th October. At the same time, large precipitation amounts have been found from the Po Plain to Sicily although the evaporative capabilities of both water basins had been switched off. The calculation of backward trajectories from the model outputs and from ECMWF analysis data suggests that the tropical African region may also have served as a moisture source. This fact is further confirmed by indirect rainfall measurements and clouds over the Sahara. The task of quantifying the role of this third moisture source is made difficult by the fact that the model produces too much moisture over the Sahara during this period compared

to ECMWF. The cause of this unusual behaviour is still unknown and further studies including the calculation of moisture flux balances are required. However, the results obtained in this study suggest a rather small direct effect upon the precipitation in the Alpine region.
Finally, a link between the evolutions of the upper-level PV streamer in each episode and the differences in the respective roles played by the Atlantic and the Mediterranean for the rainfall amounts in the Alpine region has been explored. It has been found that the upper-level precursor has been advected nearly fully extended from the eastern Atlantic to the Alpine region in the Brig flood event. Its eastern flank remained then nearly meridionally aligned throughout the event, forcing the flow impinging on the southern Alps slopes to remain southerly and favouring nearly exclusively the supply of moisture from the Mediterranean. However in the Gondo event, the PV streamer develops over western France. The evolution of its eastern flank is accompanied by a turn of the flow from southwesterly to southeasterly during the episode and favours the gradual decrease (increase) of the relevance of the Atlantic (the Mediterranean) as a moisture supplier. Moreover, the streamer developed farther south and the frontal zone on its eastern flank also extended farther equatorward compared to the Brig event. It is suggested that the presence of this baroclinic zone and of the associated jet so far to the south may have favoured the transport of moisture from tropical Africa to the Mediterranean region. Again, further case studies are required in order to better describe this possibility and to better quantify the impact of this moisture of tropical origin.
On a more general frame, the study of water transport in the atmosphere remains a very interesting and challenging topic of research due to the difficulty to model and quantify water vapor budgets and to the sometimes intricate pathways taken by moisture before falling down in the form of (heavy) rainfalls (e.g. the example of the Oder Flood, Keil et al. 1999). Possible trends in water transport routes could also constitute an indicator for climate change.

Chapter 10

Impacts on the PV Streamer Evolution

On a synoptic scale, heavy rainstorms in the southern side of the Alps are very often conditioned by the co-development of a zonally narrow and meridionally elongated upper-level PV streamer over the eastern North Atlantic. The aim of this Chapter is to study the effects of the absence of evaporation from either the Atlantic or the Mediterranean upon the genesis of the PV streamer and to further refine the study made by Massacand et al. (2001) by investigating the impacts of each water basin. The method is therefore quite different from the one used in Chapter 9. Here the differences in the flow field dynamics arising from the removal of evaporation from each water basin are studied. To this end, the results of sensitivity studies from the STREAMER09 and STREAMER10 sets of numerical experiments (e.g. Chapter 8) are investigated.

10.1 The September 1993 PV Streamer

First, a comparison with ECMWF data (e.g. Fig. E.10) shows that the control (CTRL) simulation satisfactorily simulates the streamer's evolution. There is a meridionally elongated anomaly over the eastern Atlantic, present at 00 UTC 22 September (Fig. 10.1 a) and its eastward advection over France at 00 UTC 23 September (Fig. 10.1 b) and subsequent detachment from the main PV reservoir at 00 UTC 24 September (Fig. 10.1 c) are also well represented.
No relevant difference is visible at 00 UTC 22 September on the 320 K isentropic surface between each simulation (compare Figs. 10.1 a, d, g and j). However, some decrease in the vertical structure of the negative anomaly over the mid-Atlantic in the noATL and noATLMED simulations is visible as suggested by the verti-

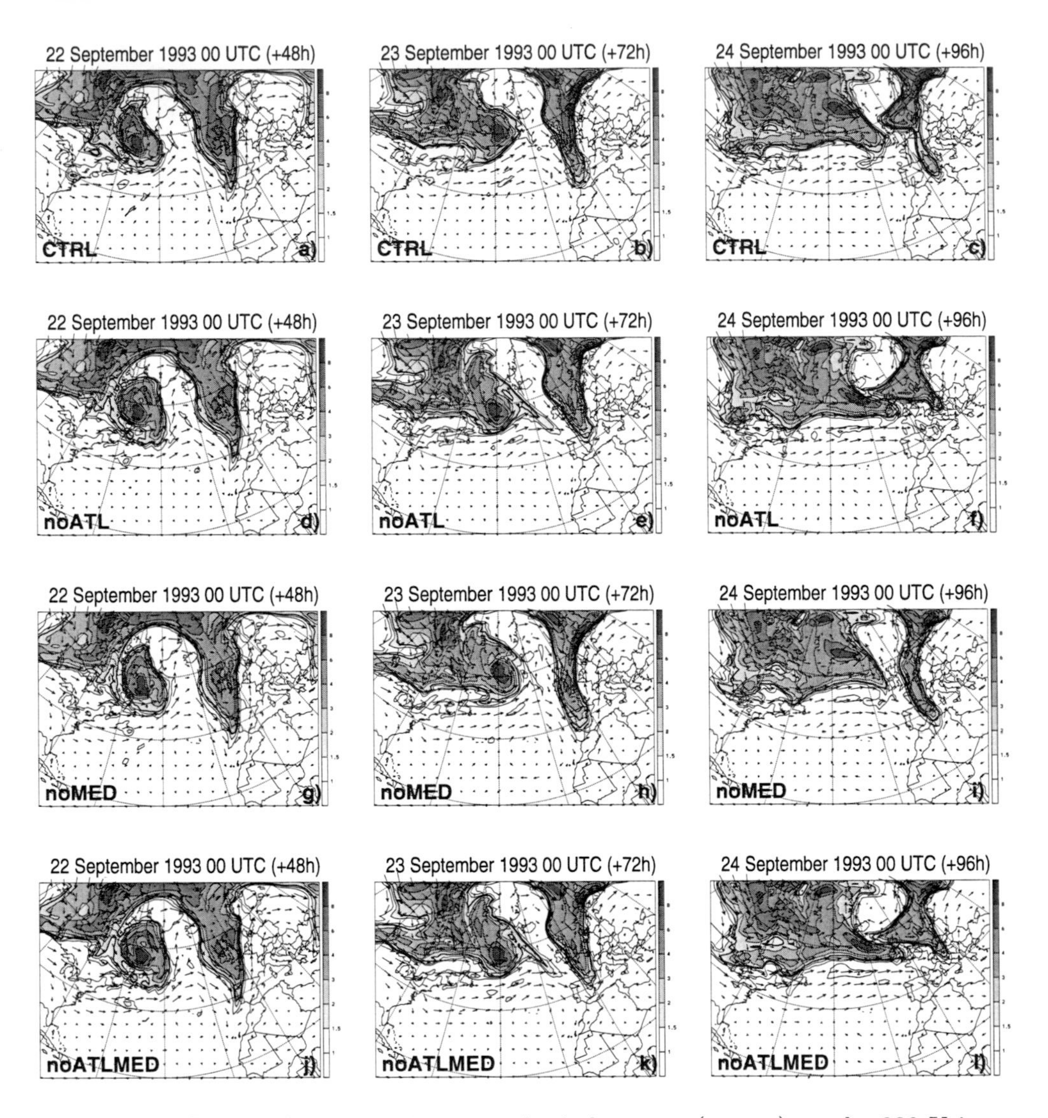

Figure 10.1: Potential vorticity in pvu and wind vectors (arrows) on the 320-K isentropic surface from the STREAMER09 set of experiments at (a, d, g and j) 00 UTC 22 September, (b, e, h and k) 00 UTC 23 September and (c, f, i, l) 00 UTC 24 September. Results from (a, b, c) the CTRL, (d, e, f) noATL, (g, h, i) noMED and (j, k, l) noATLMED simulations.

cally averaged potential vorticity in the 150-500 hPa layer (ΣPV, Figs. 10.2 a, d, g, and j).

The results for 00 UTC 23 September show discrepancies in the streamer development in the simulation without evaporation from the Atlantic (noATL and noATLMED). The streamer is shifted some 300 kilometers farther to the east

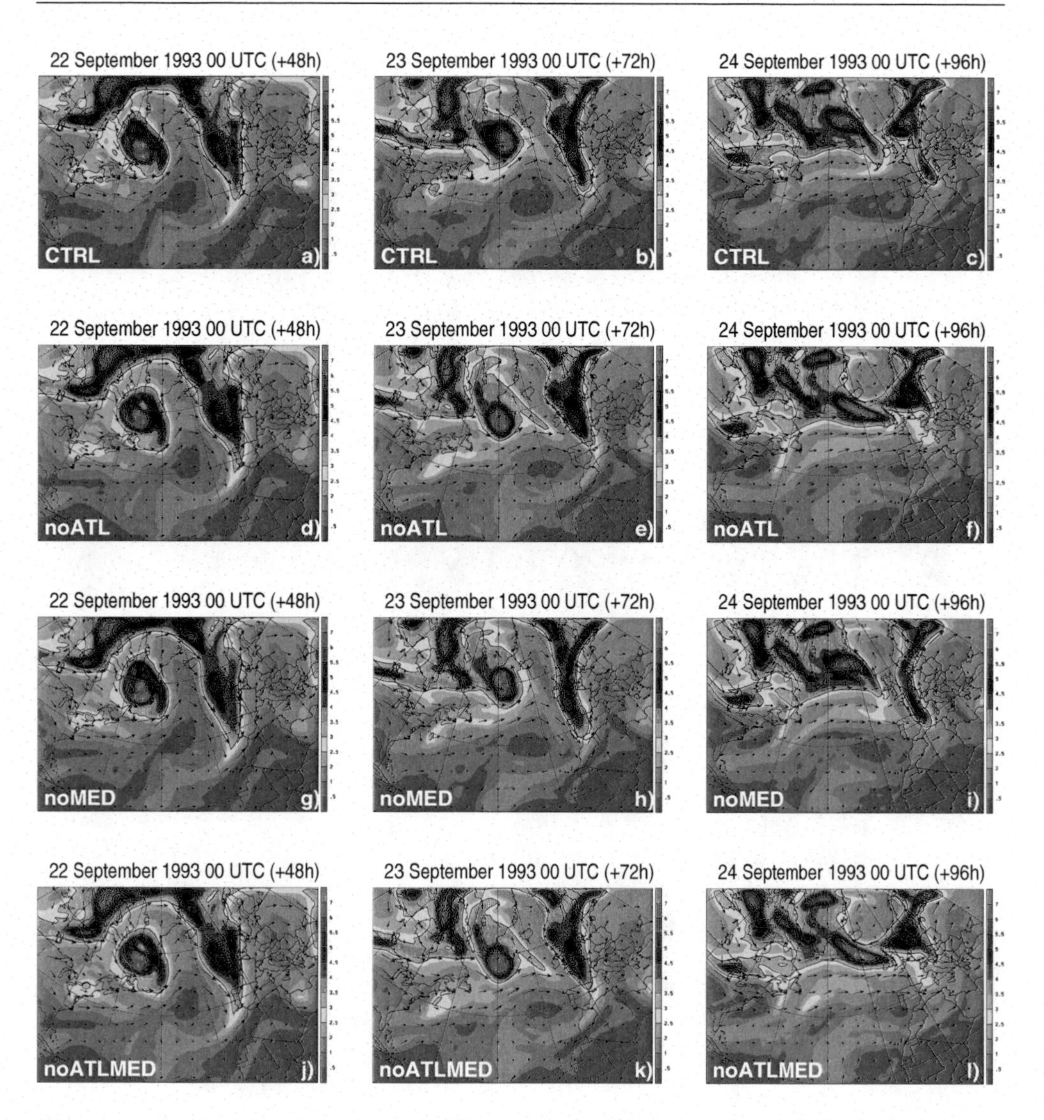

Figure 10.2: Distribution of vertically averaged PV between 150 hPa and 500 hPa (ΣPV, in pvu) and wind vectors at 300 hPa (arrows). Repartition of time and simulations is the same as Fig. 10.1.

and extends less to the south (some 500 kilometers less, e.g. Figs. 10.1 e and k) in comparison to the CTRL simulation (Fig. 10.1 b). The flow over the mid- and eastern Atlantic, as well as the flow impinging on the western flank of the streamer are slightly more zonal in the noATL and noATLMED simulations (Figs. 10.1 e and k) than in the CTRL simulation. The low PV anomaly upstream of the streamer appears to be weaker than in the CTRL simulation. No large difference is visible in the noMED simulation, except a slight broadening to the west of the

streamer (Fig. 10.1 h).
The results at 00 UTC 24 September are quite spectacular. In the simulations without evaporation from the Atlantic, the streamer extends no longer over the western Mediterranean but is located over eastern Austria (Figs. 10.1 f and l). Moreover, its meridional extension has shrunk dramatically in comparison to 00 UTC 23 September. An important consequence of this change of shape and location of the streamer is that the flow over the Alpine region has turned westerly to northwesterly at 300 hPa (Figs. 10.2 f and l), suggesting the near disappearance of any potential risk for heavy precipitation in the Alpine region. In the noMED simulation (Fig. 10.1 i), the streamer position and extension are comparable to the CTRL. However, it appears to be zonally broader and slightly deeper (Fig. 10.2 i).

10.2 The October 2000 PV Streamer

As it was the case for the September 1993 simulations, the CTRL simulation replicates quite well the evolution of the PV streamer with its development over western France at 00 UTC 12 October (Fig. 10.3 a), its southward elongation phase at 00 UTC 13 October (Fig. 10.3 b) and the separation from the main PV reservoir at 00 UTC 14 October (Fig. 10.3 c), in comparison to the ECWMF data (Fig. E.11).
No relevant difference between the simulations in the structure and position of the PV structure arises at 00 UTC 12 October.
Differences arise at 00 UTC 13 October. In the noATL and noATLMED simulations (Figs. 10.3 e and k), the streamer extends some 400 kilometers less southward and its eastern flank is located over Switzerland, resulting in a slight broadening of the streamer. The PV gradient on the 320 K isentropic surface appears to be less intense on its eastern flank and the inner structure of the streamer slightly different as large values of ΣPV are found over England (Figs. 10.4 e and k). The low-PV anomaly west of the streamer extends slightly less to the north. This feature is further confirmed by the ΣPV patterns (Figs. 10.4 e and k). In the noMED simulation, the PV streamer appears to be slightly broader on the 320 K isentropic surface (Fig. 10.3 c), but no other relevant difference is visible. The results for the noATL and noATLMED simulations at 00 UTC 14 October show large differences compared to the CTRL simulation (Figs. 10.3 f and l). A fully developed streamer is still visible but has a smaller southward extension compared to the CTRL simulation and is furthermore shifted to the east. Again, the resulting shift in the position of the eastern flank of the stratospheric intrusion is quite important for the southern Alpine rainstorm as the region of Friuli and the eastern Alps might be concerned by heavy rainfall here. No cut-off process is anticipated in the following hours as the portion of the streamer over eastern

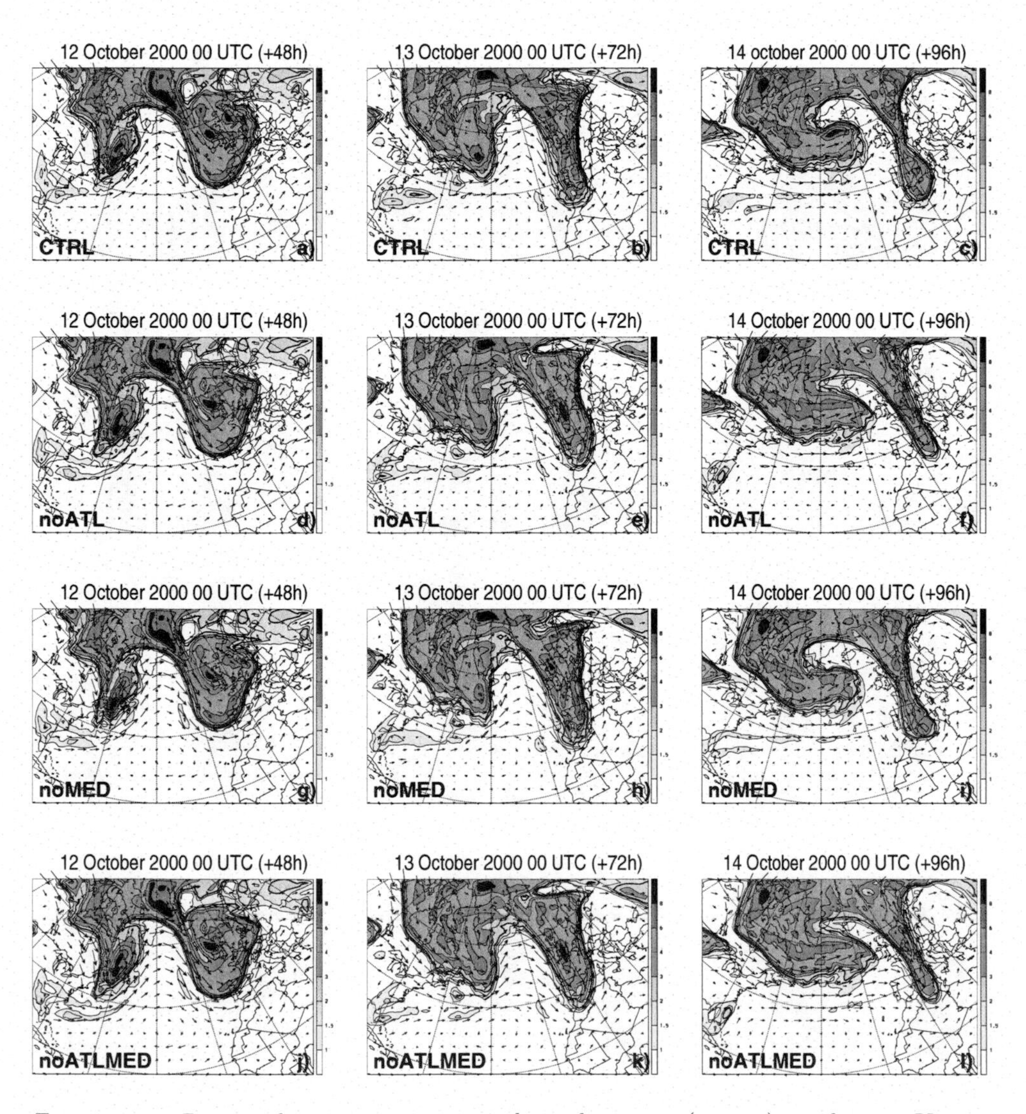

Figure 10.3: Potential vorticity in pvu and wind vectors (arrows) on the 320 K isentropic surface from the STREAMER10 set of experiments at (a, d, g and j) 00 UTC 12 October, (b, e, h and k) 00 UTC 13 October and (c, f, i, l) 00 UTC 14 October. Results from (a, b, c) the CTRL, (d, e, f) noATL, (g, h, i) noMED and (j, k, l) noATLMED simulations.

England and northern France is broader than in the CTRL simulation. This is also confirmed by the larger values of ΣPV found in this region (Figs. 10.4 f and l). The wrapping structure of the high- and low-PV anomalies over the mid-Atlantic is different as the low-PV anomaly does not extend westward over Greenland as it is the case in the CTRL simulation and the high-PV anomaly extends further

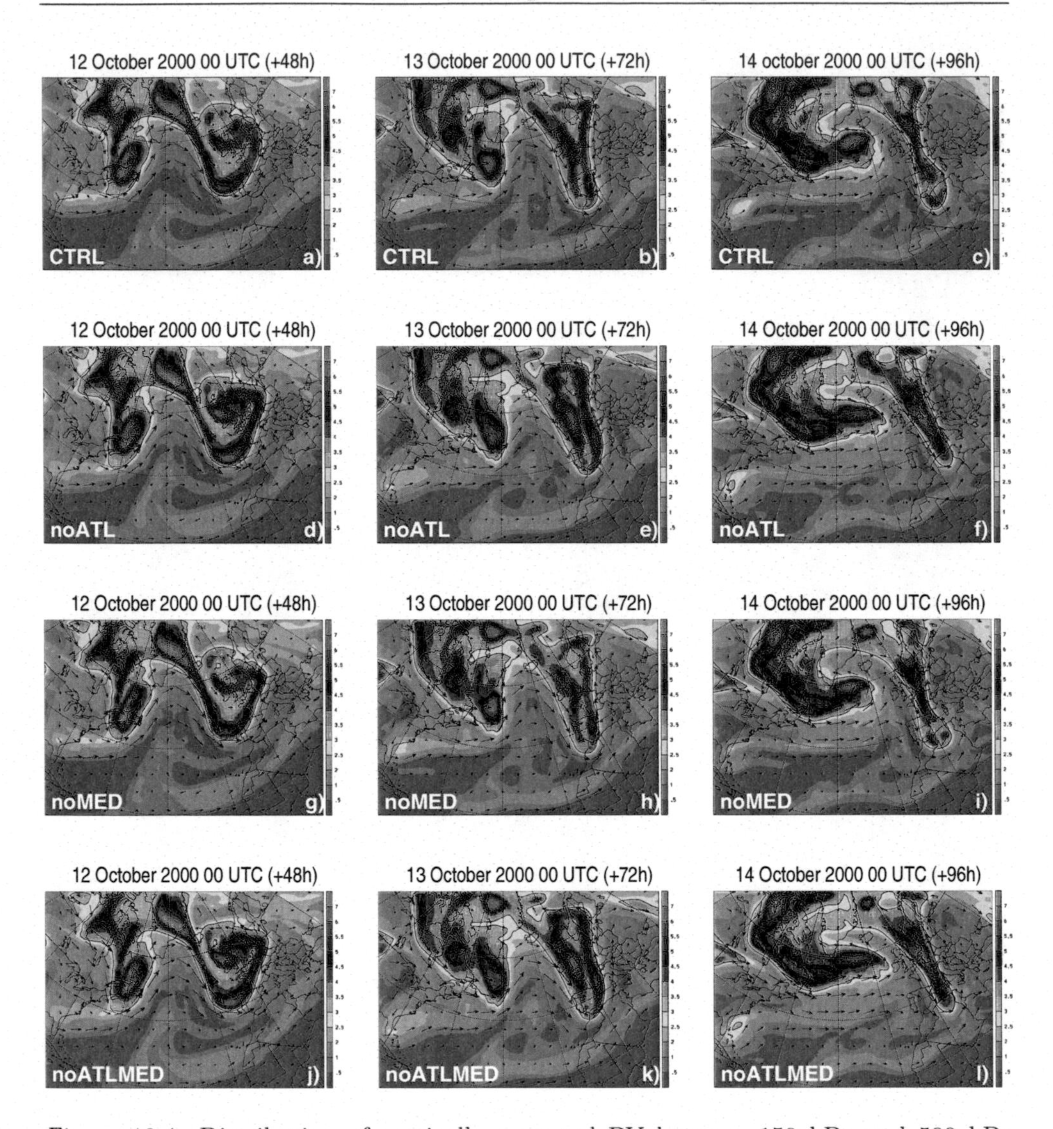

Figure 10.4: Distribution of vertically averaged PV between 150 hPa and 500 hPa (ΣPV, in pvu) and wind vectors at 300 hPa (arrows). Repartition of time and simulations is the same as Fig. 10.3.

eastward, nearly reaching the west coast of Ireland. The marked weakness of the low-PV anomaly is also visible in the vertical structure (Figs. 10.4 f and l). The noMED simulation does not show any relevant difference in the PV structure, except for a slightly broader portion of the streamer over France.

10.3 Discussion

The main interest of this Chapter is to try and isolate the individual impacts of the Atlantic and the Mediterranean as moisture suppliers on the genesis of two upper-level PV streamers that led to the occurrence of heavy precipitation on the Alpine southside. The expected influence is indirect and lies in the formation of low PV anomalies at tropopause level through tropospheric diabatic processes (Wernli 1997). Note that in this process and in a Lagrangian perspective, air parcels with low PV air are actually advected from lower levels to upper levels after having experienced an increase and a subsequent decrease of their PV value.

The results of this Chapter have shown that the Atlantic has the larger impacts on the evolution of the position and meridional extension of the PV streamer, while the Mediterranean plays no relevant role. The relevant role played by the interaction between both water basins and the variability between the two episodes underline the complexity and high non-linearity of the processes. It was also shown that the major impacts were found during the second half of the simulations.

The question under investigation in this paragraph is to know the origin of the low-PV anomaly located upstream of the PV streamer (e.g. Figs. 10.2 and 10.4). More precisely, the interest is on the thermodynamic history (adiabatic or diabatic) of the low PV air embedded within it at upper-level and on the determination of the distribution of adiabatic and diabatic PV when evaporation from the Atlantic is removed. The effects of negative anomalies upon balanced flows is quite relevant (Massacand et al. 2001 and references therein). The analysis of backward trajectories (not shown) from low PV air inside the negative anomaly shows that most of them are quasi-horizontally adiabatically advected. A second type of trajectories exhibits enhanced diabatic characteristics gained through condensation and diabatic processes (Wernli 1997). The removal of the evaporation from the Atlantic is likely to alter the characteristics of the negative PV anomaly located upstream of the streamer. Figures 10.5 and 10.6 show the vertically averaged PV and wind fields in the 150-400 hPa layer and the startiing points of 72-hour backward trajectories from the negative PV anomaly within the same layer with an initial PV value less than 0.2 pvu for the Brig and Gondo events, respectively. It appears that the removal of the Atlantic eliminates most of the trajectories having a diabatic history (i.e. an increase of the potential temperature along the trajectories larger than 10 K in the 72 hours prior to the arrival into the low-PV anomaly) in both flood events (e.g. Figs. 10.5 and 10.6 b, d and f) in comparison to the CTRL simulations (Figs. 10.5 and 10.6 a, c and e). A consequence is also visible on the vertically averaged PV in the 150-500 hPa layer (ΣPV). In the Brig case the region of very low ΣPV air located in the mid-Atlantic around 20°W/40°N in the CTRL simulation (Figs. 10.5 a, c and e) has

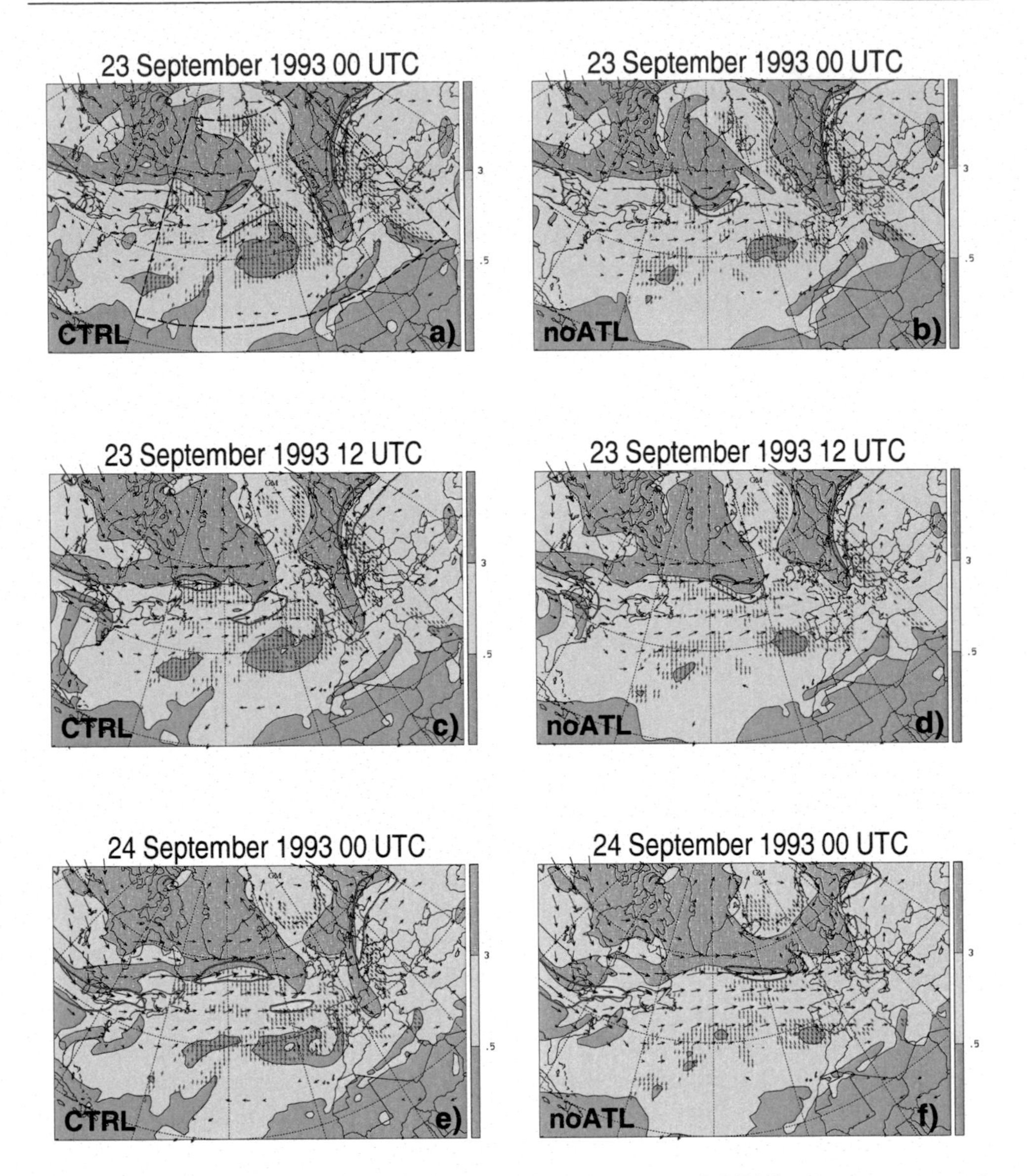

Figure 10.5: Vertically averaged PV (ΣPV, in pvu) and wind field (vectors, no vector is drawn where the vertically averaged wind speed is less than 10 ms^{-1} and green contours indicate the 30 ms^{-1} isopleth) in the 150 hPa-500 hPa layer for the Brig flood event. The red (blue) points correspond to the start of 72-hour backward trajectories ending in the 150-400 hPa layer inside the domain indicated in (a), having an initial PV value less than 0.2 pvu and experiencing a warming greater (smaller) than 10 K. Results are from the CTRL (a, c and e) and the noATL (b, d and f) simulations, on 23 September 1993 at (a,b) 00 UTC, and (c, d) 12 UTC and (e, f) at 00 UTC 24 September 1993.

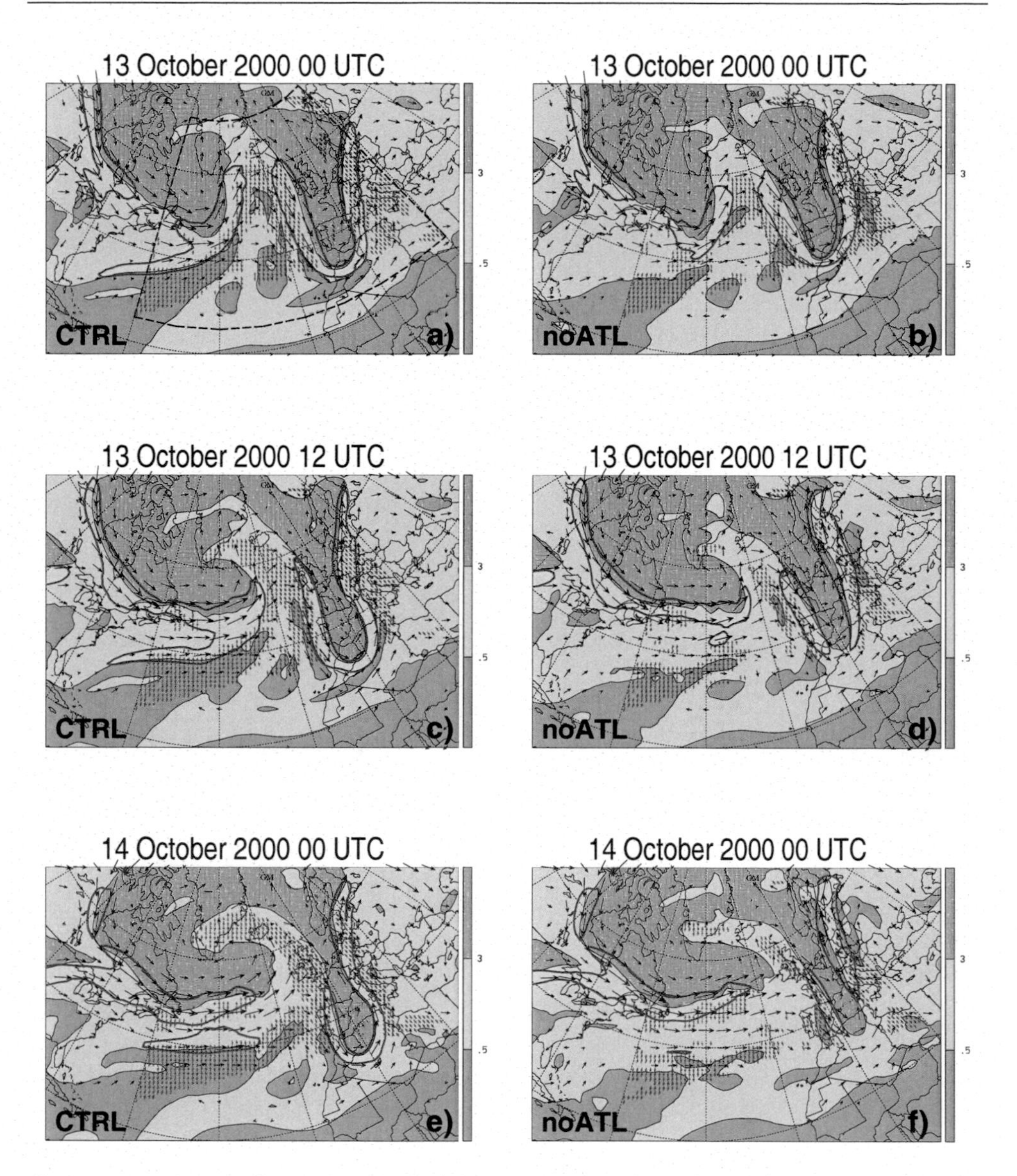

Figure 10.6: Vertically averaged PV (ΣPV, in pvu) and wind field (vectors, no vector is drawn where the vertically averaged wind speed is less than 10 ms^{-1} and green contours indicate the 30 ms^{-1} isopleth) in the 150 hPa-500 hPa layer for the Gondo flood event. The red (blue) points correspond to the start of 72-hour backward trajectories ending in the 150-400 hPa layer inside the domain indicated in (a), having an initial PV value less than 0.2 pvu and experiencing a warming greater (smaller) than 10 K. Results are from the CTRL (a, c and e) and the noATL (b, d and f) simulations, on 13 October 2000 at (a,b) 00 UTC, and (c, d) 12 UTC and (e, f) at 00 UTC 14 October 2000.

shrunk in the noATL simulation (Figs. 10.5 b, d and f). Similar results are found in the Gondo event as the regions characterized by values of ΣPV lower than 0.2 pvu present in the CTRL simulation (10.6 a, c and e) tend to shrink in the noATL simulation (10.6 b, d and f). The vanishing of these negative anomalies has a direct effect upon the characteristics of the ambient flow (magnitude and direction, structure of the upper-level jet). Finally, there is a large variability in the effects of the Atlantic upon the PV streamer from one episode to another. In the Brig flood event, the streamer recedes northward and nearly vanishes toward the end of the period. These results confirm the statement already made by Massacand et al. (2001). However, the situation is quite different in the Gondo flood event where the streamer is still present at the end of the period, although it is slightly shifted to the east and extends less southward. This suggests the relevance of other (adiabatic) actors in the genesis of the streamer, like the propagation of Rossby waves or the interaction of individual high-PV anomalies with the midlatitude PV gradient (Schwierz et al. 2004).

10.4 Summary

In this Chapter, the effects have been investigated of the Atlantic and the Mediterranean upon the genesis of two upper-level PV streamers that were precursors to two heavy precipitation episodes in the Alpine southside in September 1993 (Brig flood event) and October 2000 (Gondo flood event) by means of numerical simulations where the evaporation of each water basin has been neutralized.
The noATL and noATLMED simulations display major impacts upon the evolution of the PV streamer in both events. The streamer was advected farther downstream and exhibited a lesser equatorward elongation. Discrepancies nonetheless exist between both events. In the September event, the streamer recedes poleward by the end of the period while in the October event, the streamer is still well established over the western Mediterranean. This difference illustrates the high variability of the effects that exist from one event to the other. In the noMED simulations, the streamer tends to be zonally slightly broader but no relevant difference in the position or meridional extension has been found.
The role of the Atlantic has been further studied. On a Lagrangian perspective, it appears that the removal of the evaporation dramatically decreases the quantity of air parcels arriving at upper level in the low-PV anomaly west of the streamer after having experienced large cross-isentropic transport from the lower levels as previously suggested by Massacand et al. (2001). This in turn corresponds to a decay of very low-PV anomalies over the mid-Atlantic. This redistribution of the PV patterns directly influences the ambient flow and jet structure that will eventually impacts on the streamer evolution.

Chapter 11

Retrospective Look and Future Perspectives

This thesis has comprised two independent parts that address two specific aspects of dynamic meteorology.
The first part focussed on jet streams located in the vicinity of the tropopause. These ribbons of fast moving air are important for the aviation industry and are accompanied by salient features for the characterization of the dynamic state of the atmosphere: a vertical wind shear associated with a horizontal temperature gradient, a break in the tropopause height and strong gradients of potential vorticity on isentropic surfaces.
In order to better describe these localized regions of strong winds and to shed a new light on the study of their properties, a novel type of climatology has been developed for jet streams. This new method aims at identifying jet events through the choice of objective dynamically-based criteria. Frequency charts of jet events have been calculated that indicate *how often* jet streams occur in a specific region. Seasonal and interannual variations of the distributions of jet event patterns show the annual cycle of jet streams. The detailed information about the time and position of jet events has led to the creation of a jet event database that will eventually allow further studies.
A novel objective method has been proposed to distinguish two types of upper-level jet streams and is directly derived from the event-based jet-stream climatology. This method, in contrast to the subjective distinction between the so-called subtropical and polar-front jets, is based upon the determination of the vertical wind shear associated with the jets. To this end, two categories based upon baroclinic properties have been introduced: the shallow-layer and deep-layer baroclinicity categories. These two categories permit a direct partition of jet events.
A search has been made for particular hemispheric flow configurations where single or double jets occur and their respective zonal distributions have been de-

termined. Single jets are more frequent in the Northern Hemisphere where they are mainly found in the western Pacific and western Atlantic sectors. Single and double jets are equally distributed in the Southern Hemisphere except between the eastern Indian Ocean and the mid-Southern Pacific where double jets clearly dominate. The mean atmospheric state accompanying temporal and spatial coherent single- or double-jet flow configurations in specific regions has been studied with the composite technique. Differencies exist from one hemisphere to the other in the mean position and baroclinic properties of the jets for each type of flow.
The methods developed to isolate jet events and multiple jets imply some limitations linked to the direction of the jet. Information are offset about the direction of the jet during the analysis as only the norm of the wind speed is calculated. In the climatologies of Chapter 4 the jet event frequency patterns appear as zonal bands. Nevertheless, cases can occur where the jet is actually meridionally oriented (i.e. in the case of meridionally elongated troughs). The calculation of the climatologies then tends to 'smooth' the patterns and give them their zonal appearance. This fact has to be remembered when the jet event climatologies are analyzed. The lack of information about the direction can also have an indirect influence on the detection of multiple jets. Suppose a Rossby wave breaking pattern with a ridge/trough pattern nearly zonally oriented. In this case, the multiple-jet detection technique will identify three jets (a westerly jet at the edge of the main PV reservoir and an easterly/westerly pair of jets on each side of the trough) although it is actually the same jet that is associated with the Rossby wave pattern. The coupling of this method with an identification technique of PV contours on isentropic surfaces (e.g. Zillig 2002) could be an asset for the identification of multiple jets.

The second part has been devoted to the study of heavy precipitation on the Alpine southside that are accompanied at upper-level by the presence of a precursor in the form of a meridionally elongated zonally narrow intrusion of stratospheric air (PV streamer). The focus has been the assessment of the relative roles played by Atlantic and Mediterranean as moisture sources for the flood episodes of September 1993 and October 2000. To this end, numerical simulations have been conducted in which the evaporation of one or both water basins has been suppressed. The motivation has been twofold. First the relative impacts upon the intensity of the precipitation in the Alpine region have been studied. Second, the sensitivity of the development of the upper-level PV precursor to the evaporation from each water basin has been investigated. This last point constitutes a further refinement in the study of the causal chain of physical processes proposed by Massacand et al. (2001) that contribute to the streamer's generation.
In the first set of experiments, the study examined the response to changes to the moisture distributions. This study of the impacts of the Atlantic and the Mediterranean as moisture sources has shown significant differences between the

two episodes. In the case of the September 1993 flood event, nearly all the rainfall can be related to the evaporation from the Mediterranean. Conversely, in the first half of the October 2000 flood event, both water basins have contributed as moisture sources to the rainfall on the Alpine southside. In the second half the influence of the Atlantic has decreased relative to that of the Mediterranean. Calculations of Lagrangian trajectories suggest the possible involvement of moisture originating in the tropical region of Africa. This difference in the moisture source involvement for the two heavy precipitation episodes has been put in relation to differences in the characteristics of the accompanying upper-level PV streamer.
In the second set of experiments, changes in the dynamics have been investigated. The highly non-linear nature of the development of the upper-level PV precursor is illustrated by the episode-to-episode variability of the relative influences of each water basin. The Atlantic plays a relevant role in the meridional extension and zonal positioning of the streamer in both episodes. However, the influence of the Atlantic appears to be essential in the September 1993 event as the streamer totally disappears at the end of the simulation without evaporation from the Atlantic. A similar result for this event has been found by Massacand et al. (2001) who used dry simulations. Conversely, the Atlantic has a weaker influence in the October 2000 event as a streamer is still present though shifted to the east and meridionally less extended. These results are relevant as the presence or absence of the PV streamer or the differences in its position and structure have direct consequences on heavy precipitation risks and on the determination of regions affected by these extreme events. It is noteworthy that the influence of the Mediterranean upon the development of the streamer appears to be marginal in both episodes.
Some limitations of the results of Part II arise from the attribution of the relative roles of each water basin. In this Part, highly non-linear processes are tackled. The factor separation method used give insight to individual contributions but one has to be careful with the interpretation of the results. This applies especially for the interaction term whose patterns are sometimes difficult to explain. A single factor can for instance influence many other factors that are not analyzed with the method. These factors may then also interact and as the effect is not proportional to the cause in a nonlinear system, the interpretation of the interaction becomes delicate.

Future Perspectives

The results obtained in both parts of this thesis point to several possible items for future investigations.
The jet event database enables case studies of single or double jets. From a synoptic point of view particular extreme events (such as the winter storm 'Lothar'

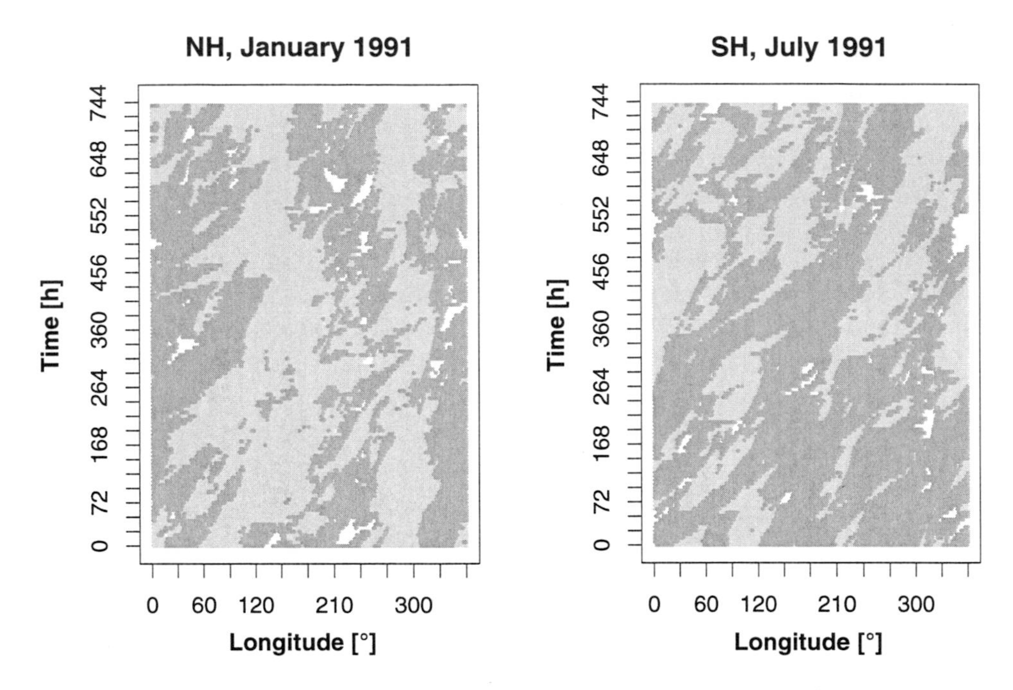

Figure 11.1: Hovmöller diagrams depicting single-jet (light grey points) and double-jet (dark grey points) events for the Northern Hemisphere (January 1991, left panel) and the Southern Hemisphere (July 1991, right panel). The white regions correspond to events of the $\mathcal{J}_R$ category (e.g. Eq. 6.1, Section 6.1).

in December 1999, e.g Wernli et al. 2002) can be retrieved by means of weather data and the associated evolution and configuration of the jet streams can be studied. On a more general point, cyclogenesis characteristics in single- and double-jet configurations can be tackled by studying individual cases in a first step and then generalize the results with a composite study of cyclone life cycles associated with each type of jet configuration in a second step. This would supply material for a direct verification based on real weather data of the conceptual hypothesis proposed by Shapiro et al. (1999) of the implications of the positions of two jets on the resulting wind shear profil relevant for the type of cyclone life cycle (see Fig. 3.4 in Section 3.4).

The database is helpful in retrieving individual episodes of temporally and spatially coherent single- and double-jet episodes in any region of both hemispheres. In a more large-scale perspective, questions arise on how single-jet flow configurations evolve into double-jet ones and vice-versa. To this end, Hovmöller diagrams showing the zonal distribution of single and double jets as a function of time (Fig. 11.1) can be useful. They can shed light on the behaviour of individual episodes, whether they are stationary or transient. Furthermore, conditions preceding or following the occurrence of single or double jets in a specific region can

be assessed by the (composite) study of the couple of days preceding or following the episode. This would give new insights to the characteristics of the global circulation.
Nowadays, climate change has become a major issue in atmospheric sciences. It is tempting to try and determine a fingerprint of climate change in the occurrence frequencies of jet events and in particular to see whether any long-term trend is identifiable in the occurrence of single- or double-jet episodes. However, this event-based jet analysis encompasses only the 15-year period of the ERA15 data set. Therefore the use of the newly available ERA40 data set that encompasses the period from mid-1957 to 2001 would enable a more consistent analysis of time series.
From a more fundamental point of view, jets are salient fluid dynamic features. Many unknowns remain despite many studies that have tackled the evolution of eddies into zonal jets (Rhines 1975) and compared the global circulations on Jupiter (Williams 1979a) and on Earth (Williams 1979b). It would be interesting to better describe the genesis of jets in a fluid dynamic perspective, especially to study the range of parameters in planetary circulations by means of GCMs as more and more space probes are launched to observe and study the characteristics of the Planets of the Solar system.

The determination of the moisture sources involved in heavy precipitation episodes in the Alpine region is actually only a part of a much larger theme, i.e. the transport of moisture in the troposphere. The method used in this thesis made use of a complex numerical simulation type where a 'pre-drying' of the atmosphere overlaying each water basin was first implemented before running the case-study simulation experiments. An alternative type of simulation would involve the use of a passive tracer related to moisture (Harald Sodemann, personal communication). This method would then study the evolution of the tracer in a single simulation of the event.
In this thesis, only two episodes of heavy precipitation in the Alpine southside characterized by the occurrence of an elongated PV streamer at upper level have been studied. In order to generalize the results found here, the study of more rainstorm episodes on the southern Alpine slopes characterized by an upper-level PV streamer is required. However not all PV streamers over western Europe lead to extreme rain events (Zenklusen 2004). The determination of the moisture sources can then be an asset to better understand the atmospheric conditions (moisture flux, temperature, stability) under which Alpine severe rainstorms occur.

Appendix A

Appendix to Chapter 4

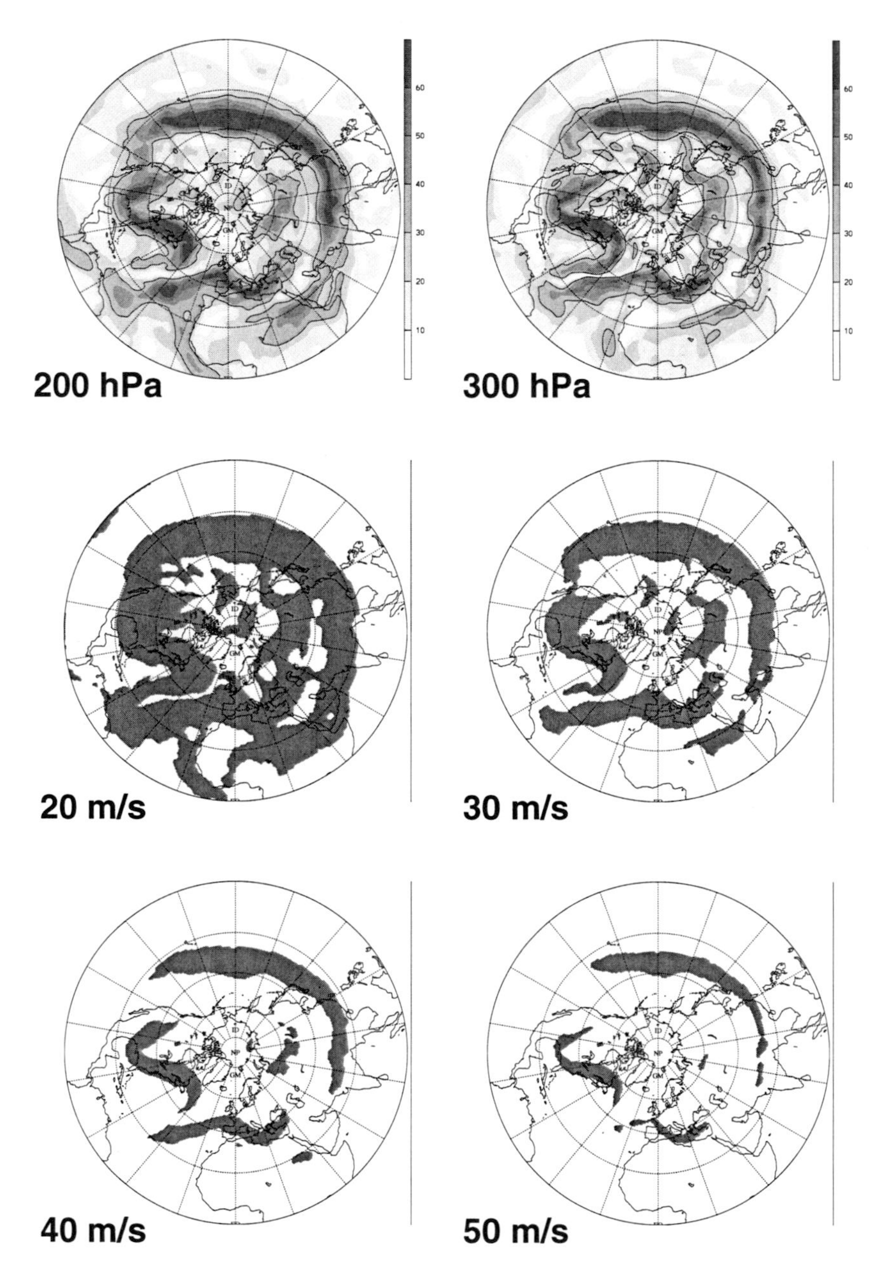

Figure A.1: Illustration of the sensitivity of the identification of jet events to the wind velocity threshold. The upper panels show the wind velocity in ms^{-1} at (left) 200 hPa and (right) 300 hPa. The four other panels show jet events identified by the method for speed thresholds criteria of 20, 30, 40 and 50 ms^{-1}.

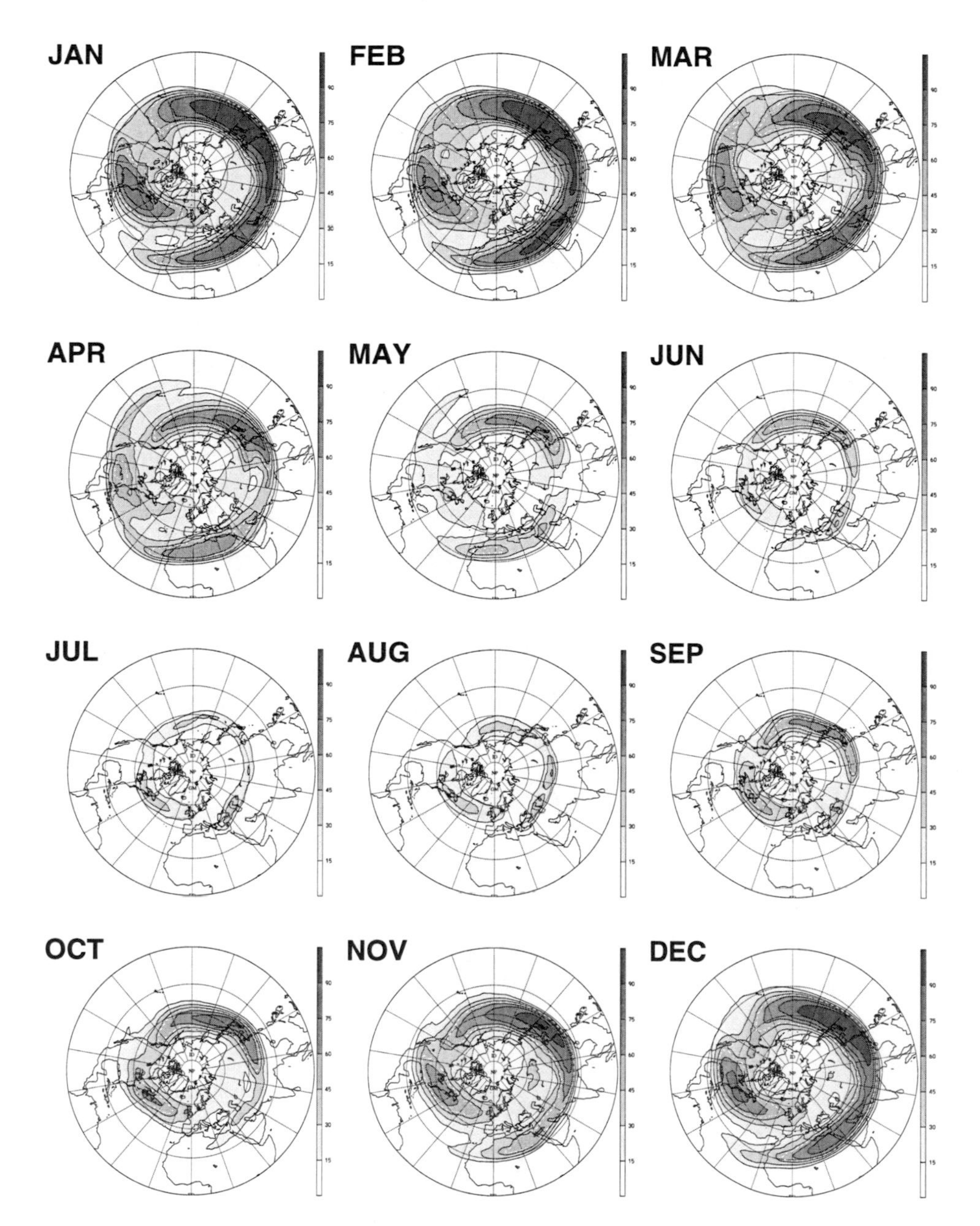

Figure A.2: Monthly climatological distribution of the jet events for the Northern Hemisphere. Values in %, the white dashed line indicates 50%.

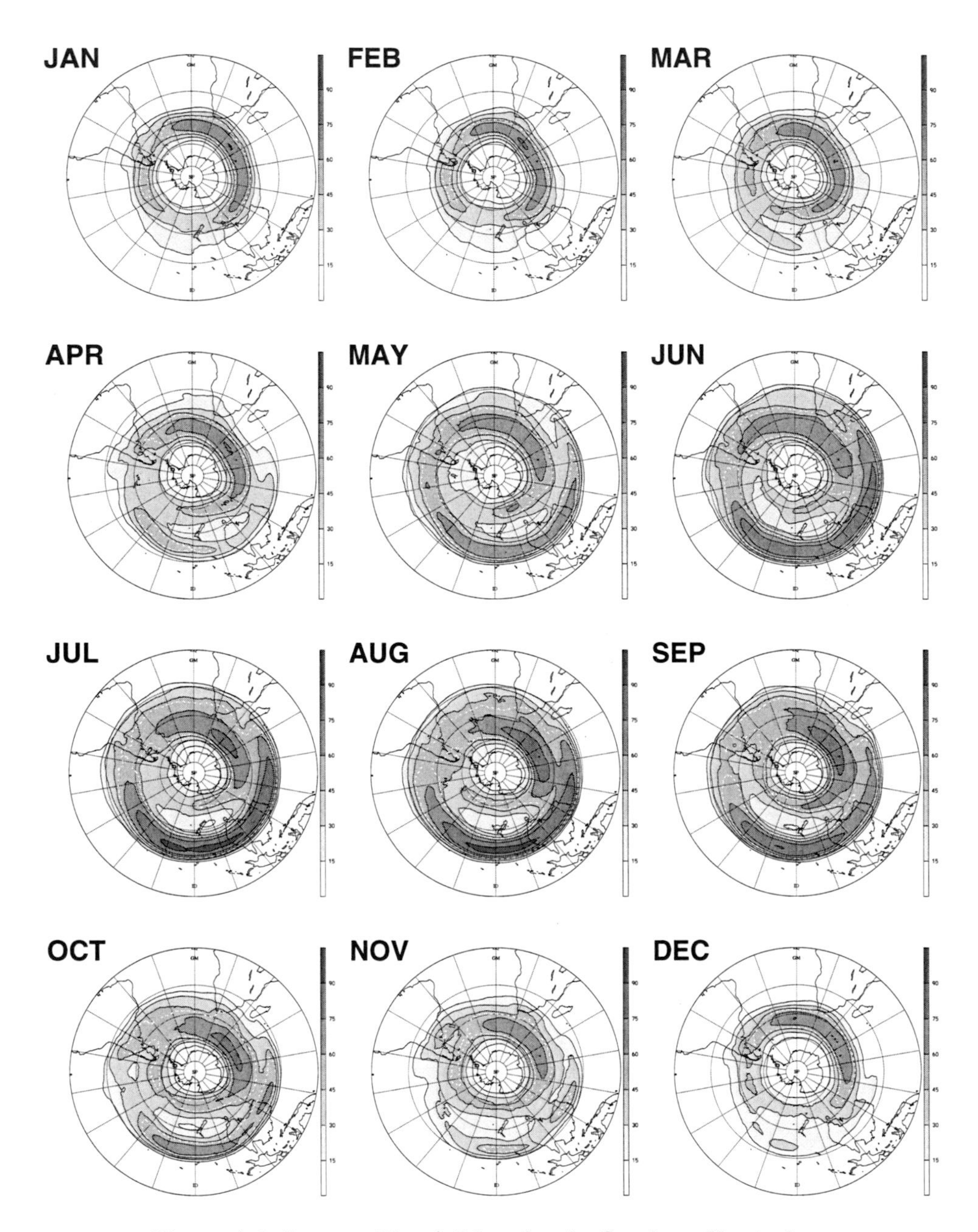

Figure A.3: Same as Fig. A.2 but for the Southern Hemisphere.

Appendix B

Appendix to Chapter 5

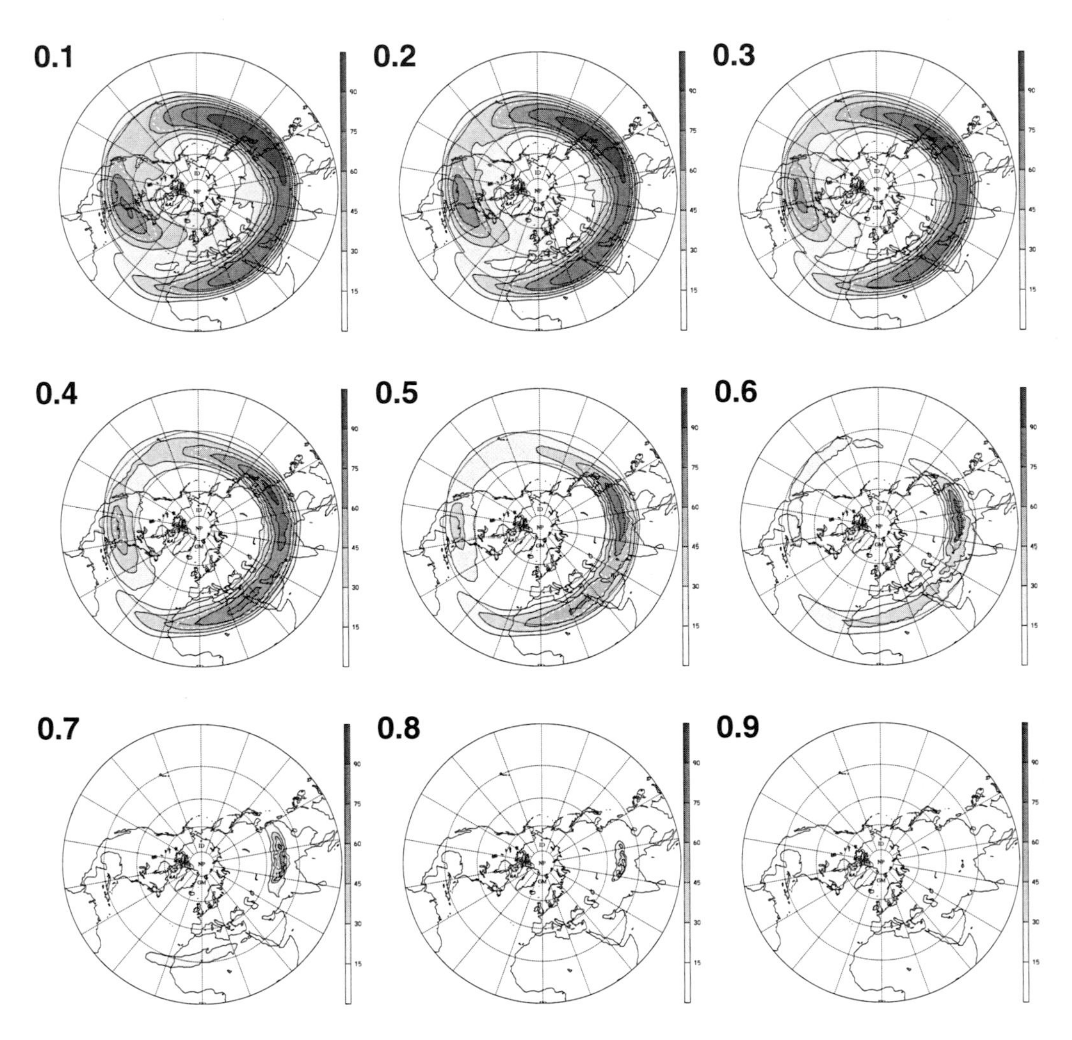

Figure B.1: Jet event frequencies in % of the SL category for different threshold values for Δv_{rel} for winter in the Northern Hemisphere. The dashed white line indicates 50%.

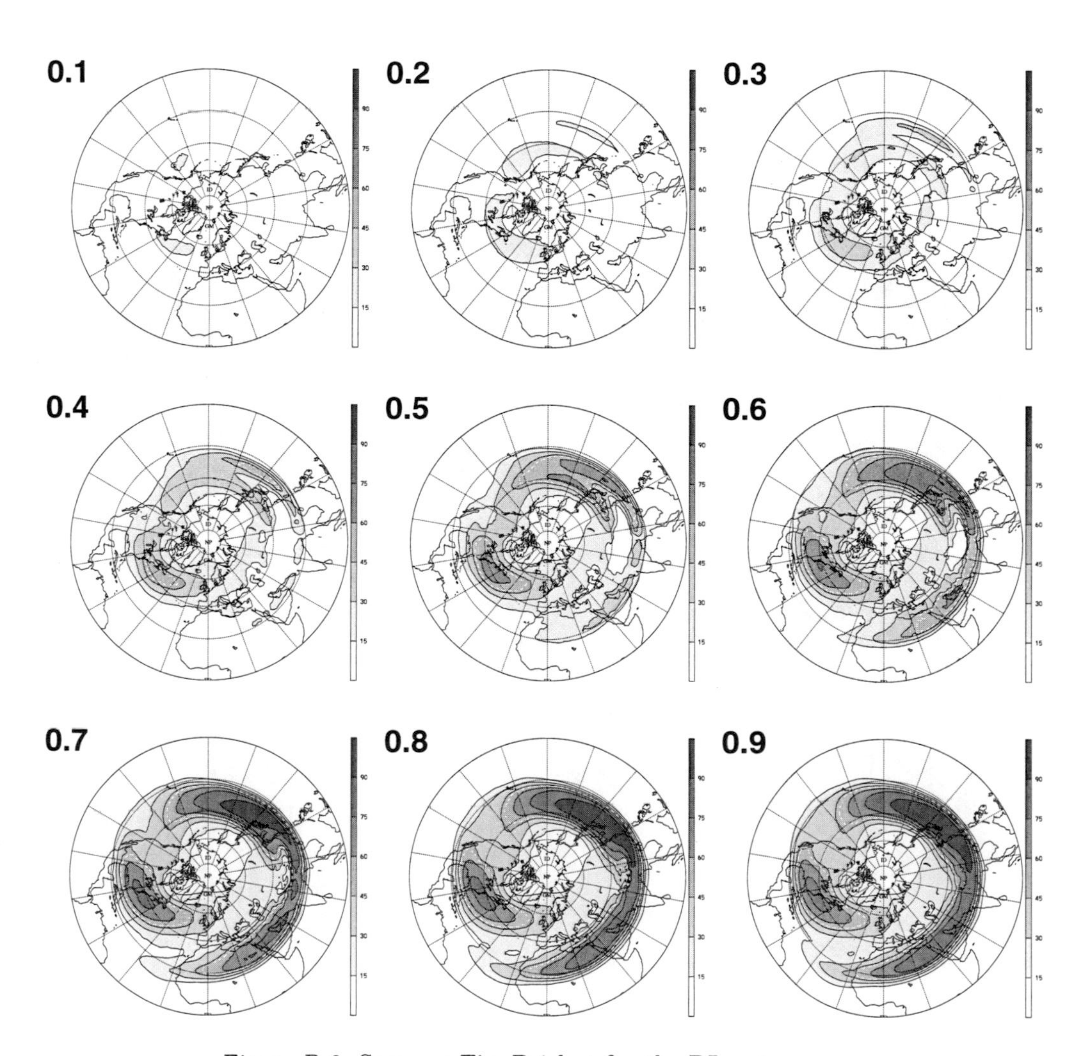

Figure B.2: Same as Fig. B.1 but for the DL category.

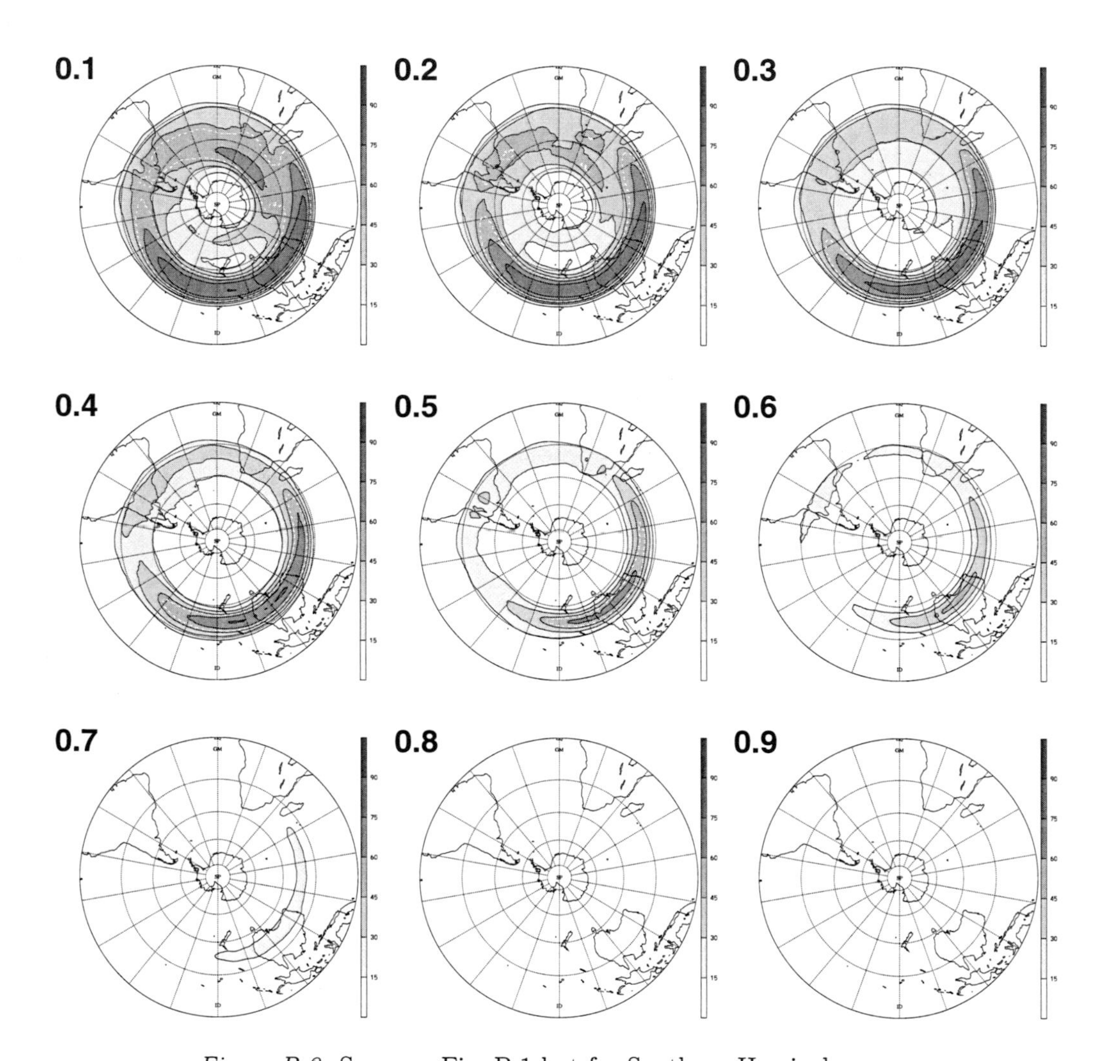

Figure B.3: Same as Fig. B.1 but for Southern Hemisphere.

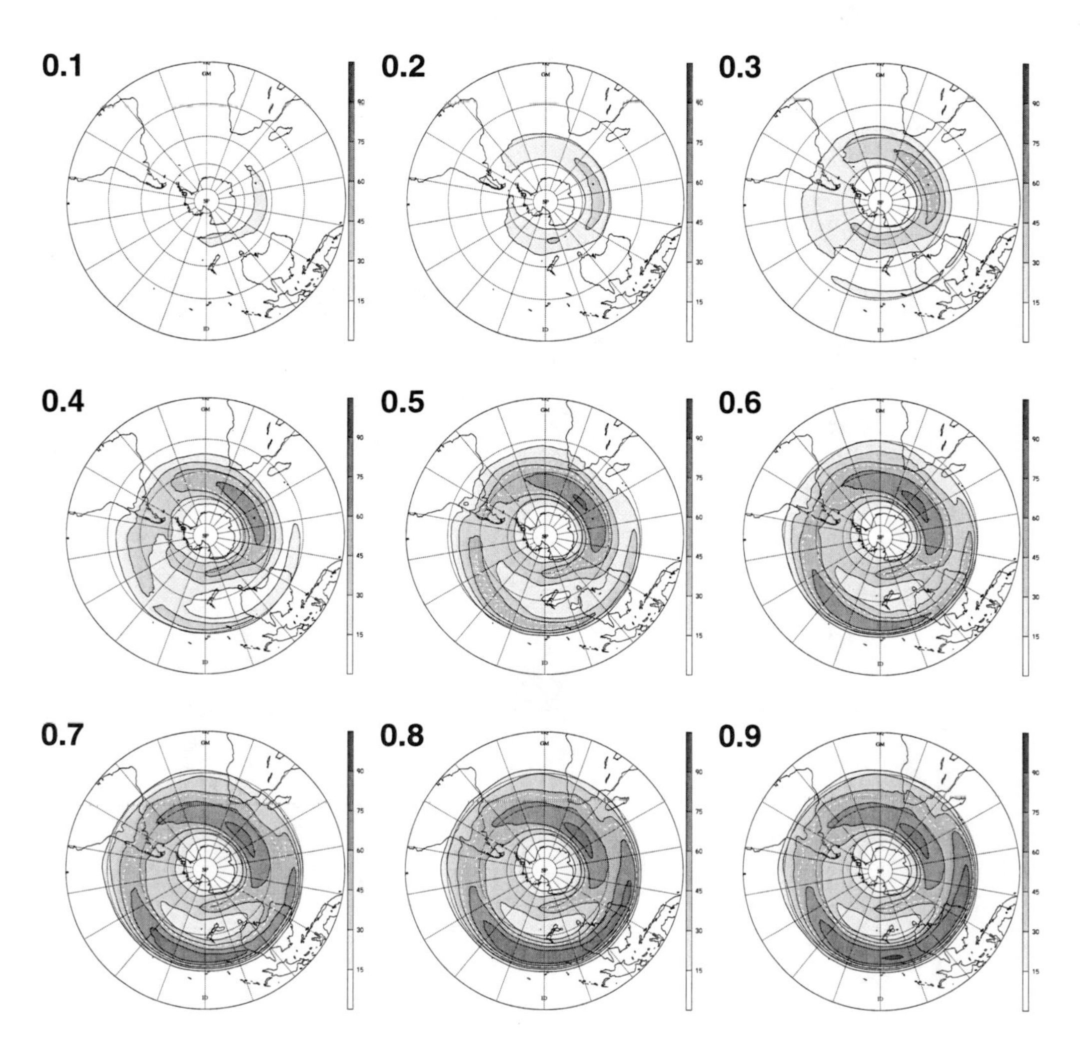

Figure B.4: Same as Fig. B.3 but for the DL category.

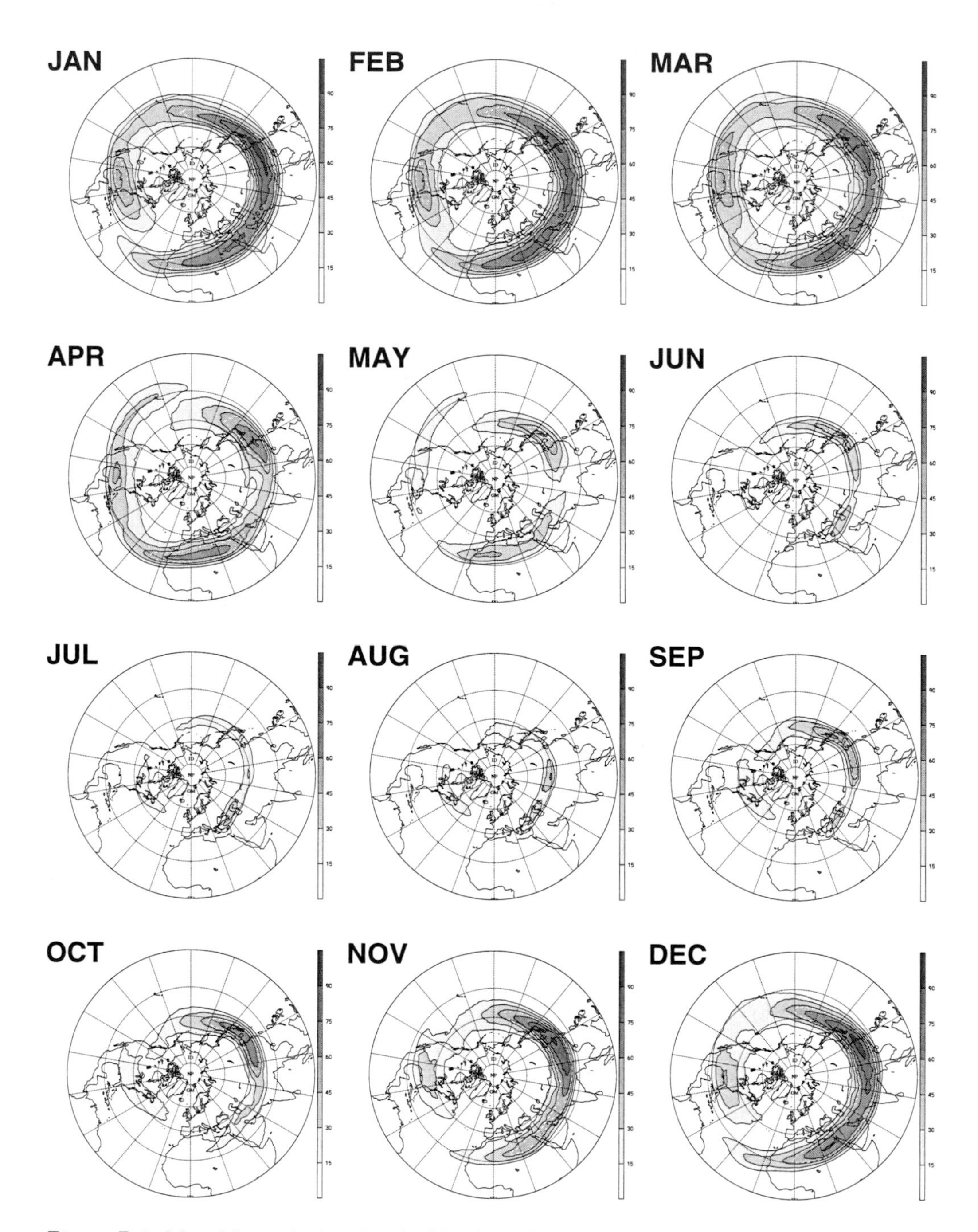

Figure B.5: Monthly evolution for the Northern Hemisphere of the jet event frequency (in %) of the SL category for a threshold value for Δv_{rel} of 0.4. The dashed white line indicates 50%.

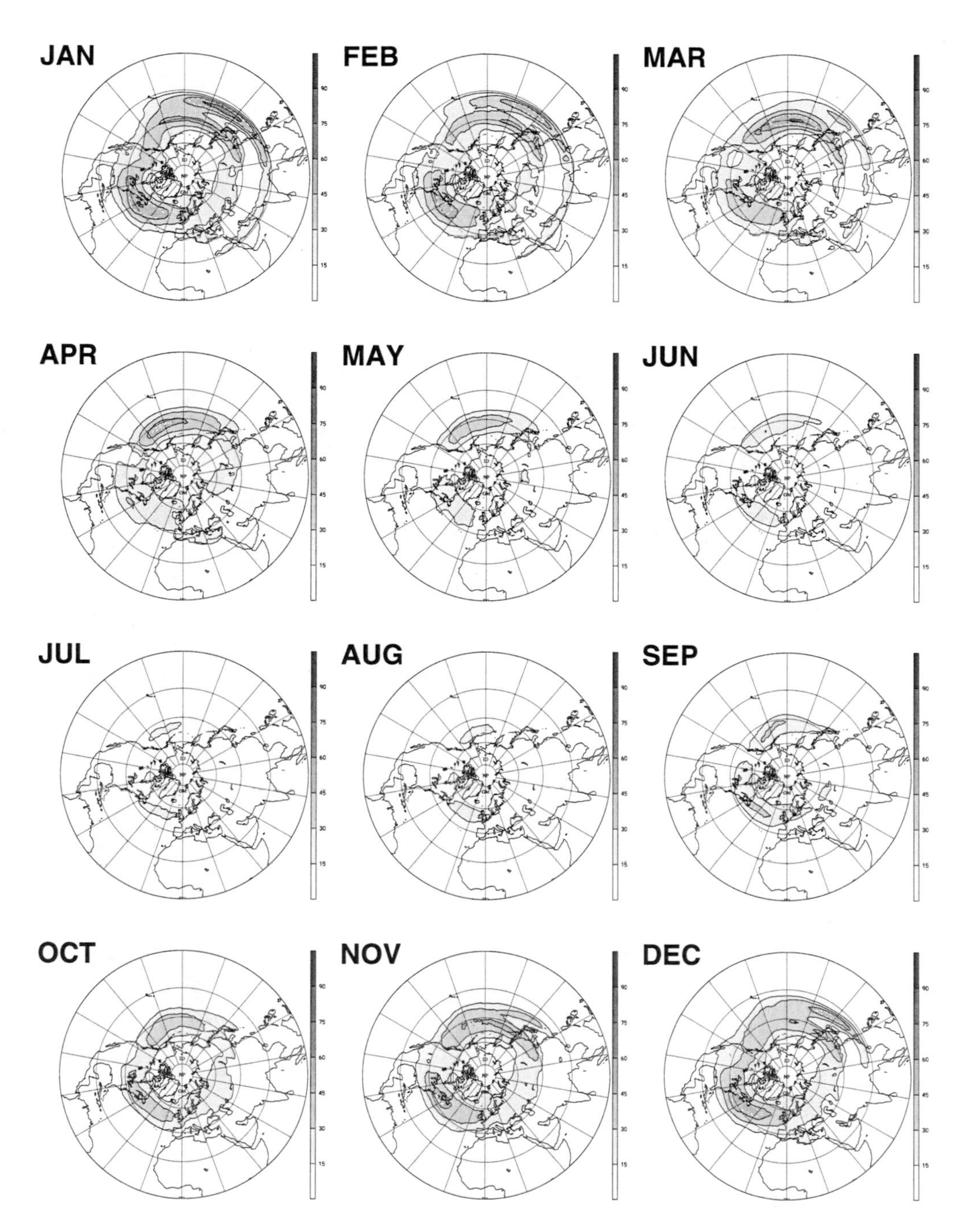

Figure B.6: Same as Fig. B.5 but for the DL category.

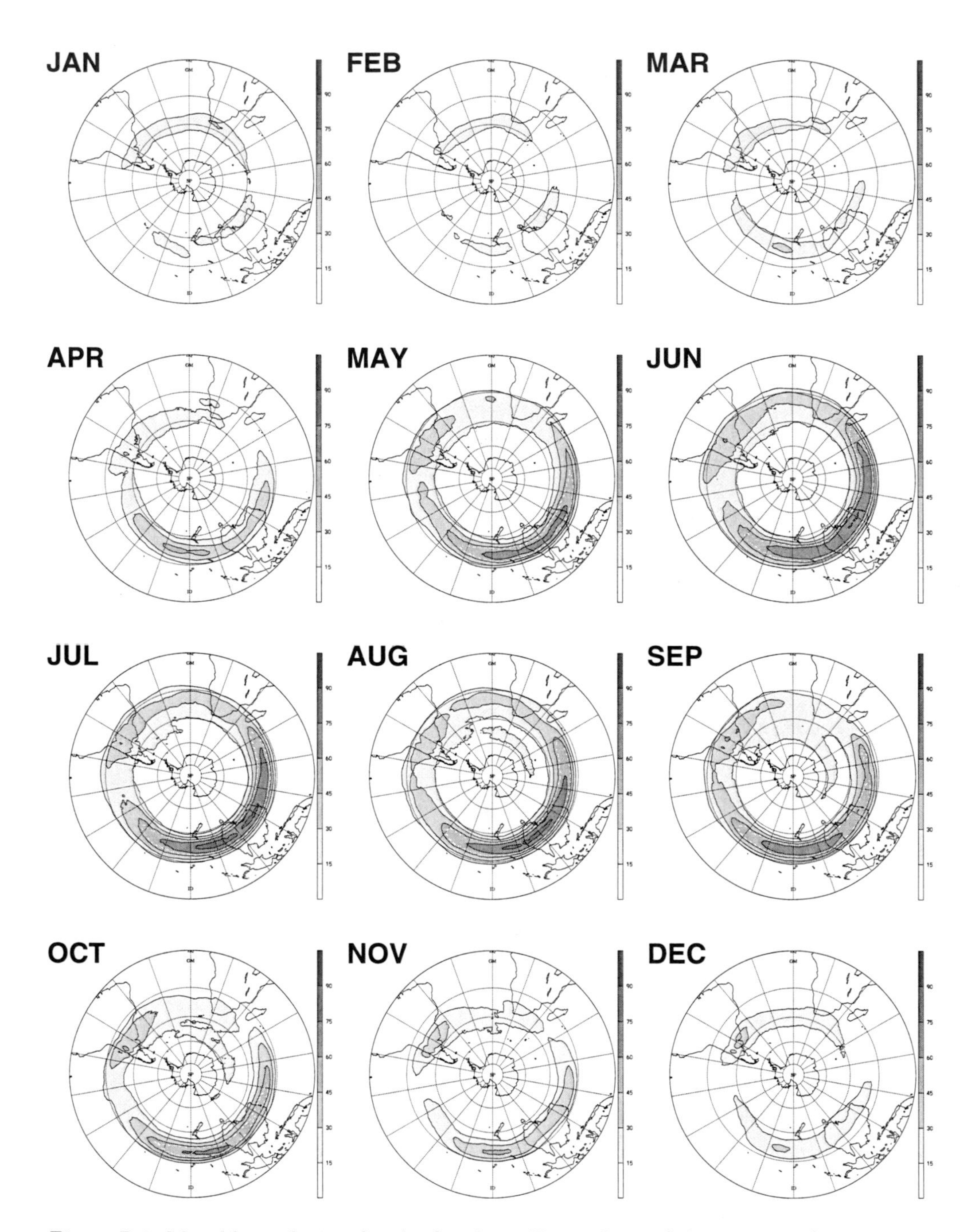

Figure B.7: Monthly evolution for the Southern Hemisphere of the jet event frequency (in %) of the SL category for a threshold value for Δv_{rel} of 0.4. The dashed white line indicates 50%.

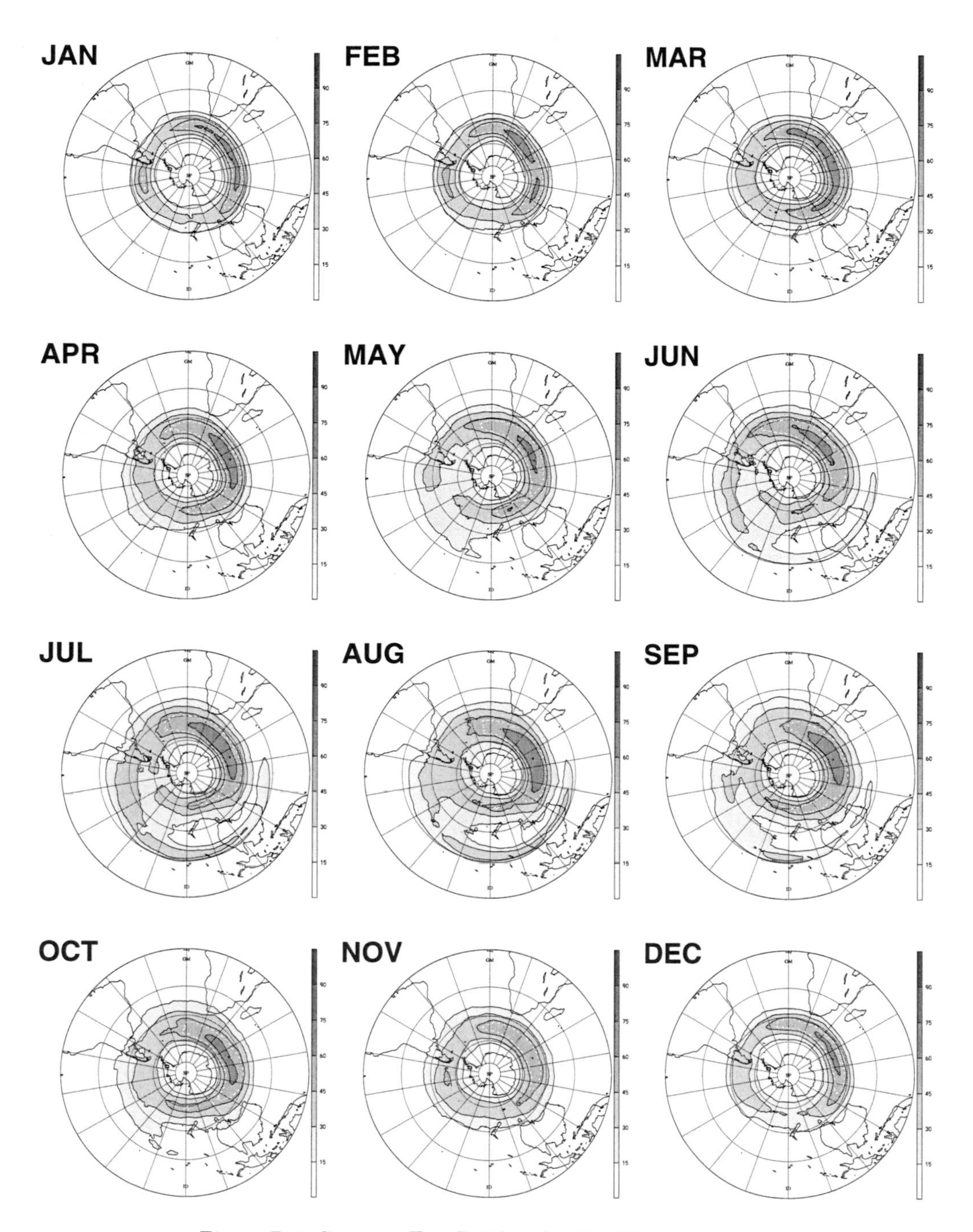

Figure B.8: Same as Fig. B.7 but for the DL category.

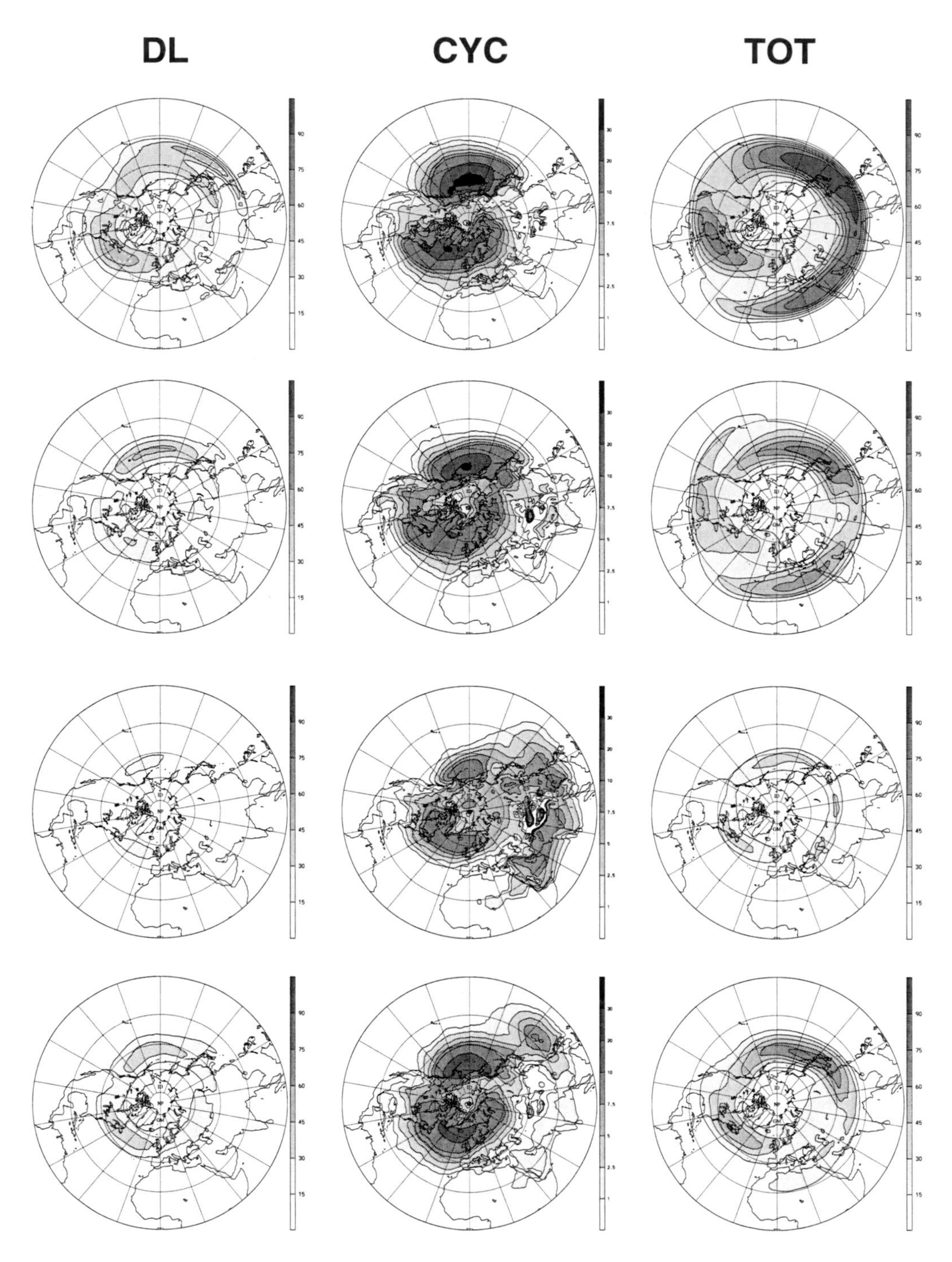

Figure B.9: Comparison for the Northern Hemisphere of moving cyclone climatologies (middle column) with the DL (left column) and the total (right column) jet event climatologies. All scales are in %. First row is for DJF, second one for MAM, third one for JJA and last row for SON.

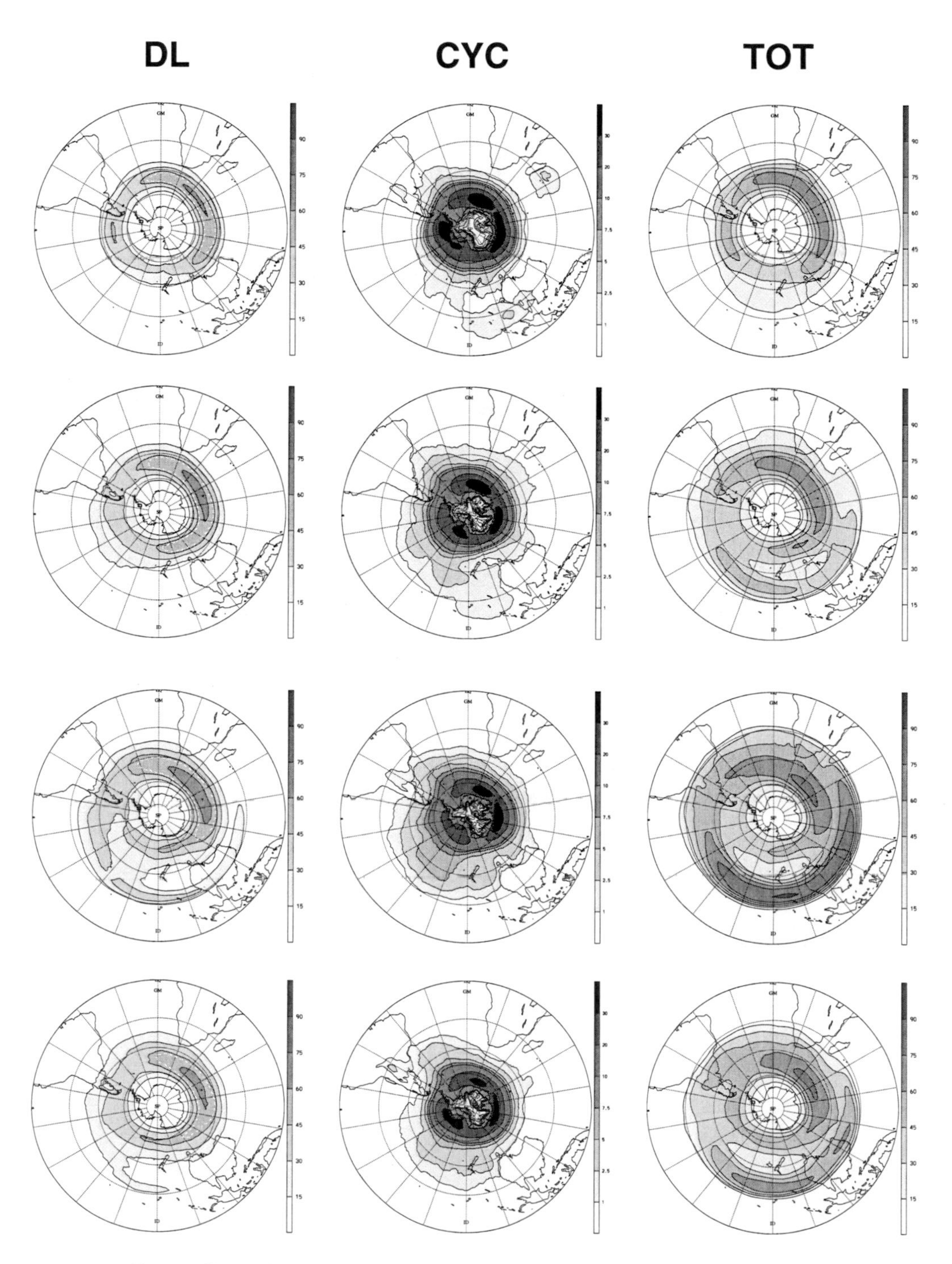

Figure B.10: Same as Fig. B.9, but for the Southern Hemisphere.

Appendix C

Appendix to Chapter 6

NH	December	January	February	Total
Single-Jets	111	26	54	**191**
Double-Jets	216	310	303	**829**
ERA15	1860	1860	1696	**5416**

Table C.1: Monthly distribution of the respective number of time instances found to match the zonal length criterion of the single- and double-jet events for winter in the zonal interval defined in Section 6.3.1 (30°W-40°E) in the Northern Hemisphere (NH). The total number of time instances of the boreal winter months of the whole ERA15 period is also given for comparison.

SH	June	July	August	Total
Single-Jets	30	43	31	**104**
Double-Jets	377	451	382	**1210**
ERA15	1800	1860	1860	**5520**

Table C.2: Same as Table C.1 but for the single- and double-jet events in winter in the zonal interval defined in Section 6.3.1 (130°E-210°E) in the Southern Hemisphere (SH). The total number of time instances of the austral winter months of the whole ERA15 period is also given for comparison.

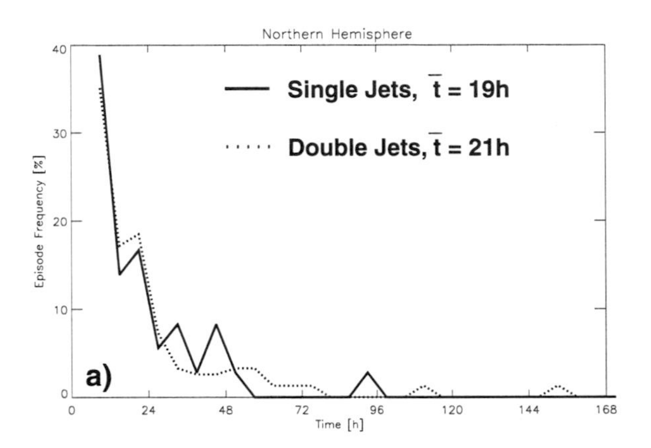

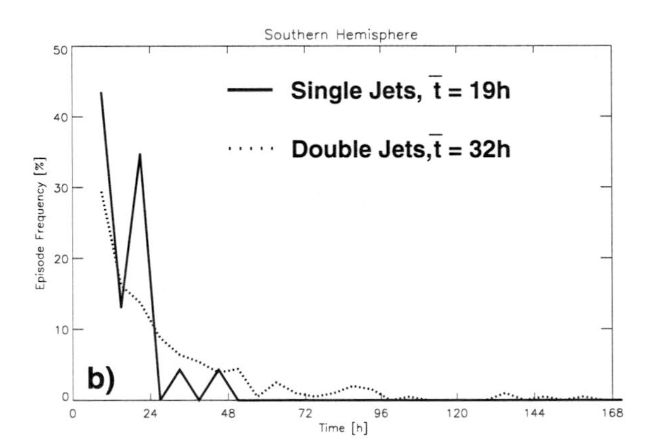

Figure C.1: Distribution of the normalized number of single- and double-jet episodes following their relative durations. The solid (dotted) line depicts single-jet (double-jet) episodes for (a) the Northern Hemisphere and (b) the Southern Hemisphere. The mean durations of the episodes of each jet configuration are given. Episodes with durations less than 6 hours have been omitted. See text for more details.

Appendix D

Appendix to Chapter 8

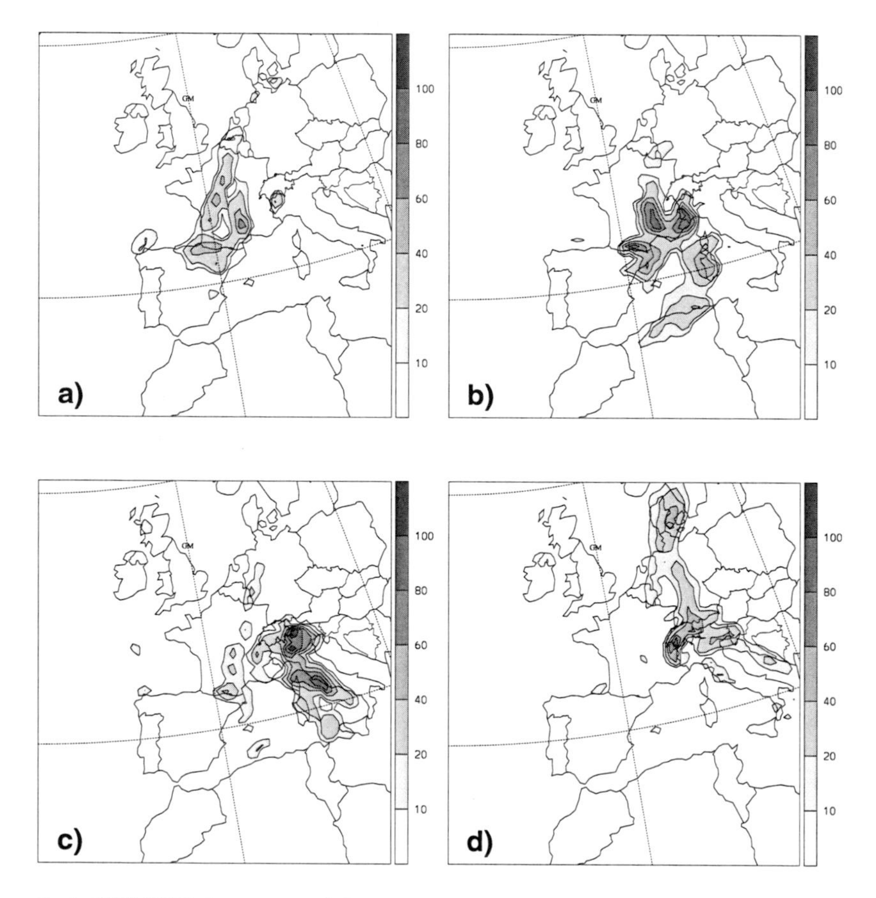

Figure D.1: ECMWF forecasts of the 24-hour accumulated precipitation in millimeters for the September 1993 heavy precipitation episode. Results are for (a) 22 September, (b) 23 September, (c) 24 September and (d) 25 September.

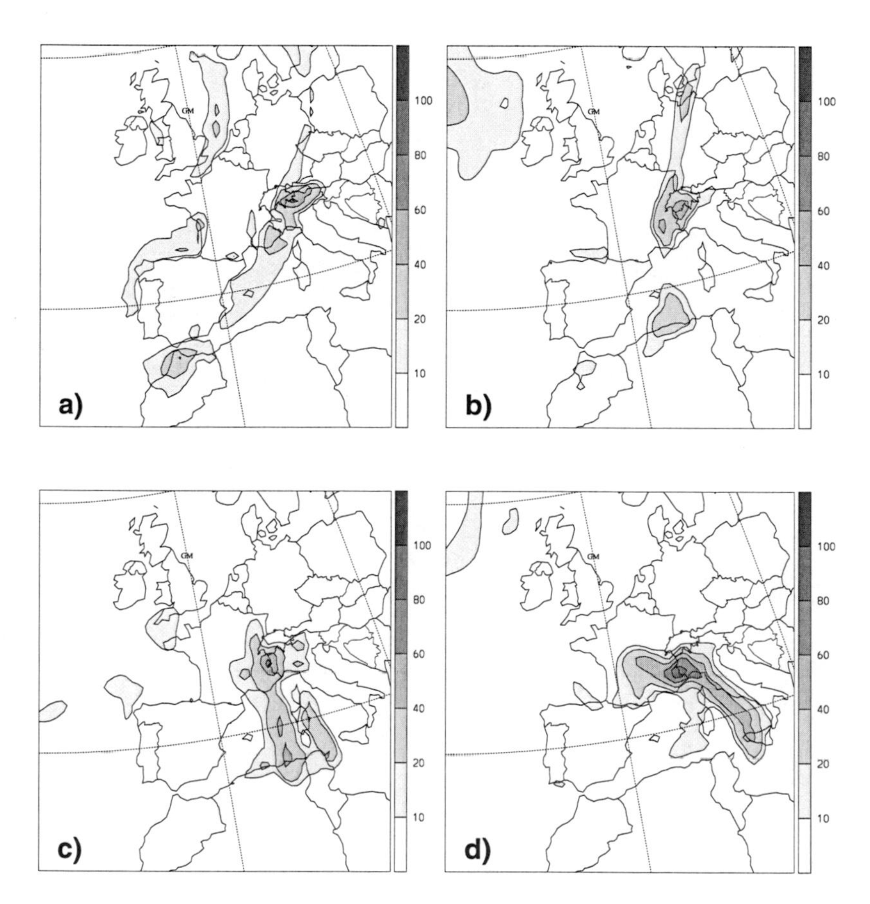

Figure D.2: ECMWF forecasts of the 24-hour accumulated precipitation in millimeters for the October 2000 heavy precipitation episode. Results are for (a) 12 October, (b) 13 October, (c) 14 October and (d) 15 October.

Appendix E

Appendix to Chapter 9

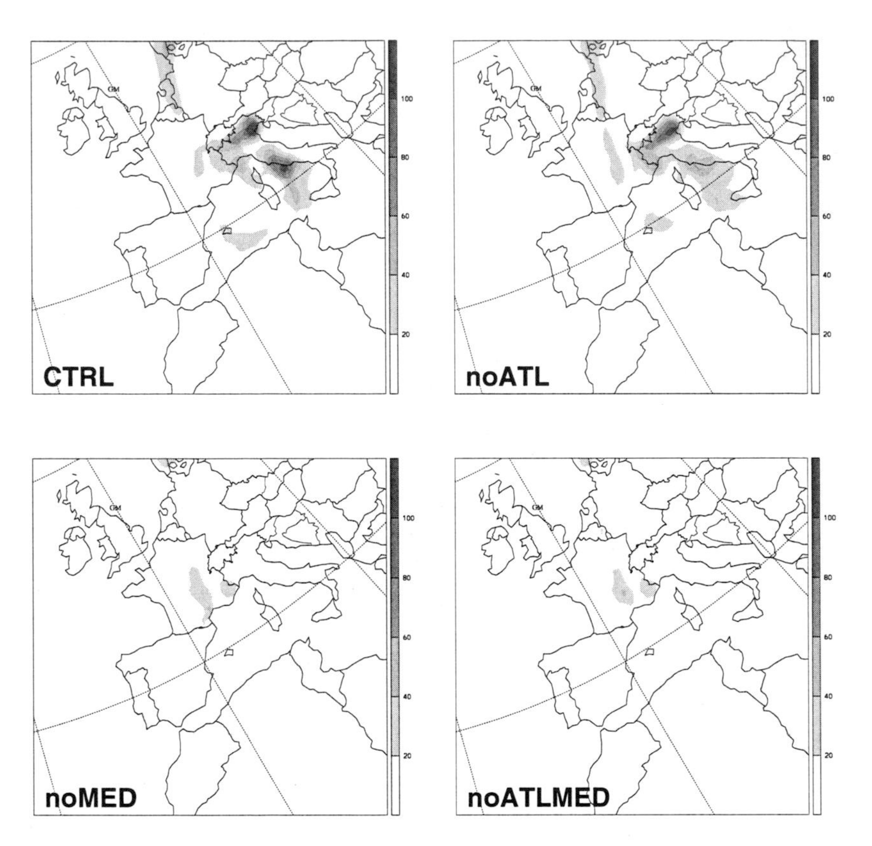

Figure E.1: 24-hour accumulated precipitation in millimeters on 24 September obtained from the BRIG2 experiment. CTRL, noATL, noMED and noATLMED refer to the control simulation and to the simulations without evaporation from the Atlantic, the Mediterranean and from both the Atlantic and the Mediterranean, respectively.

The vertically integrated (between 800 and 1000 hPa) horizontal moisture flux $\bar{WV}$ presented on Fig. E.2 through Fig. E.7 is calculated the following way:

$$\bar{WV} = \int_{1000}^{800} Q \cdot (u^2 + v^2)^{1/2} dp \tag{E.1}$$

where Q is the total specific humidity, u and v the zonal and meridional wind components, respectively.

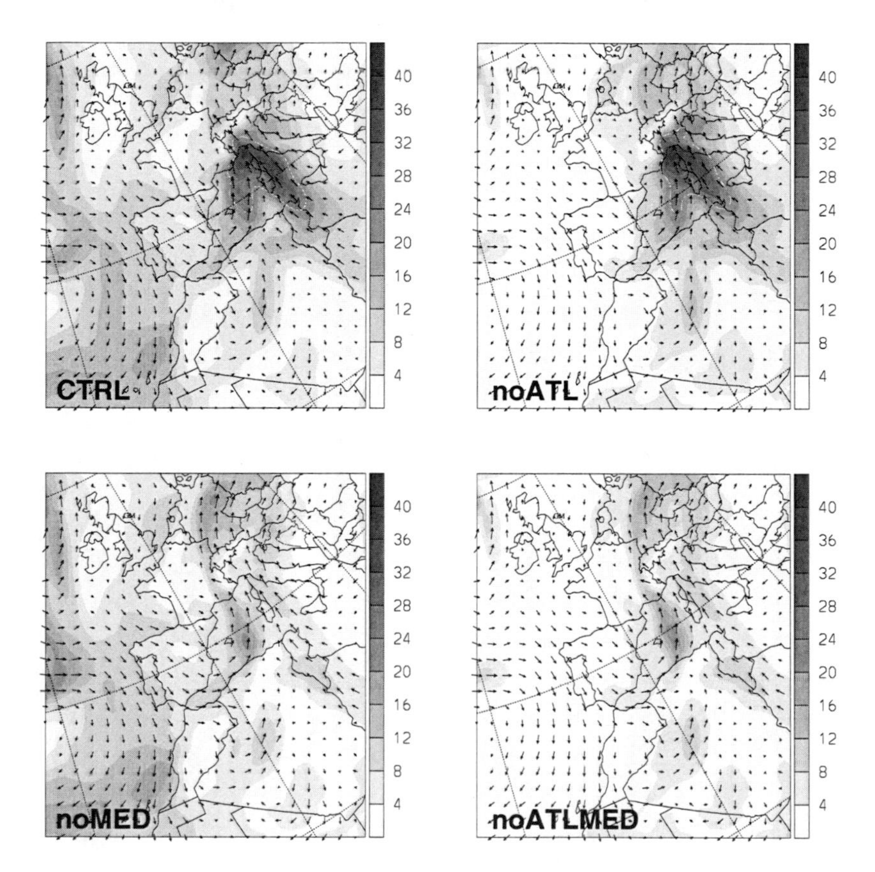

Figure E.2: Horizontal moisture fluxes vertically integrated from 1000 hPa to 800 hPa (shaded, unit: $kg\ [H_2O]/s^3$) and wind vectors at 850 hPa. The white dashdotted line outlines the 24 $kg\ [H_2O]/s^3$ isoline. Results for 12 UTC on 24 September, from the BRIG2 experiment.

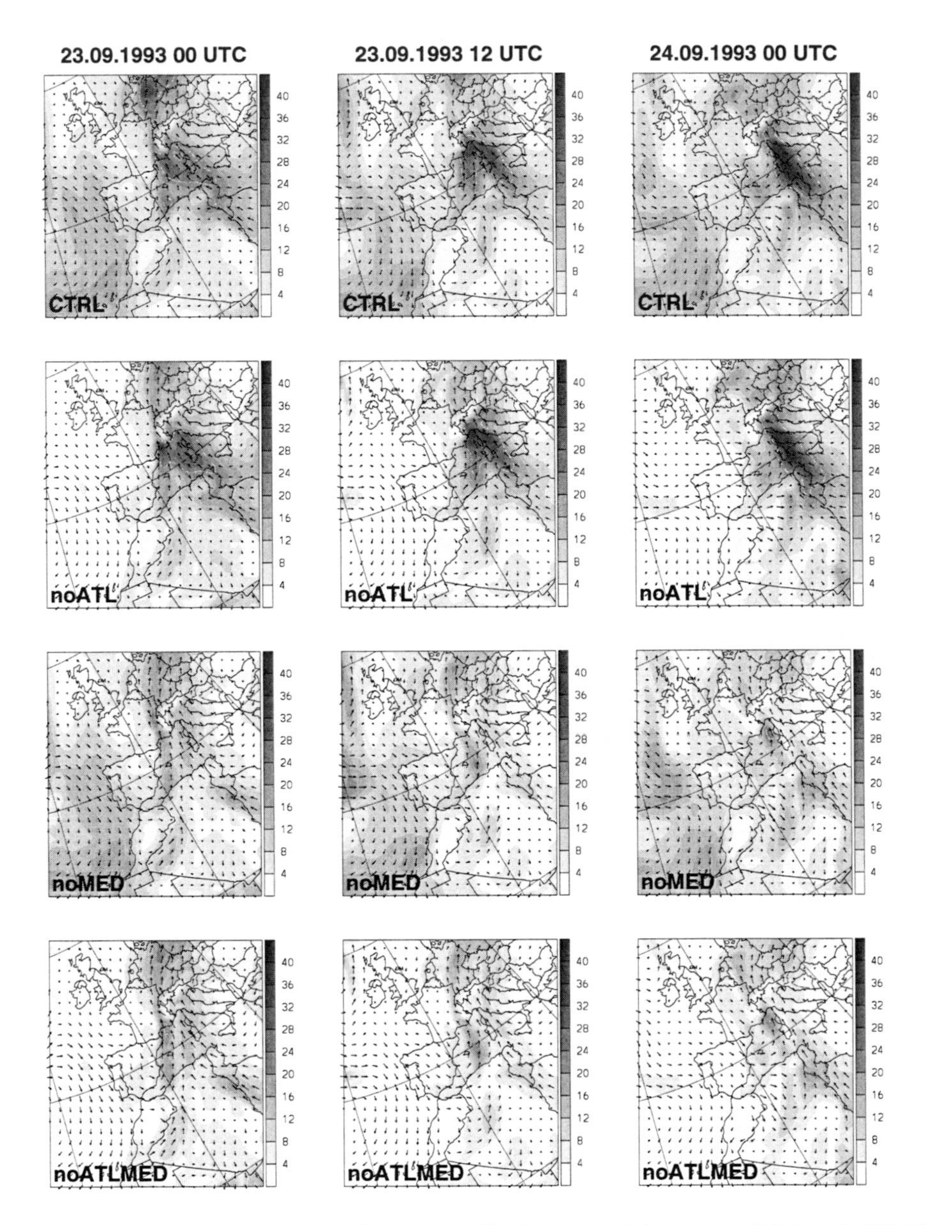

Figure E.3: Horizontal moisture flux vertically integrated between 1000 and 800 hPa (shaded, in kg H_2O/s^3, the 24 kg H_2O/s^3 isoline outlined with a white dashdotted line) and wind vectors at 850 hPa. Results from BRIG1 for 23.09.1993 00 UTC (+24h, left-hand panels), 23.09.1993 12 UTC (+36h, middle row panels) and for 24.09.1993 00 UTC (+48h, right-hand panels). CTRL, noATL, noMED and noATLMED refer to the control simulation and the simulations without evaporation from the Atlantic, the Mediterranean and from both the Atlantic and the Mediterranean. See text for more details.

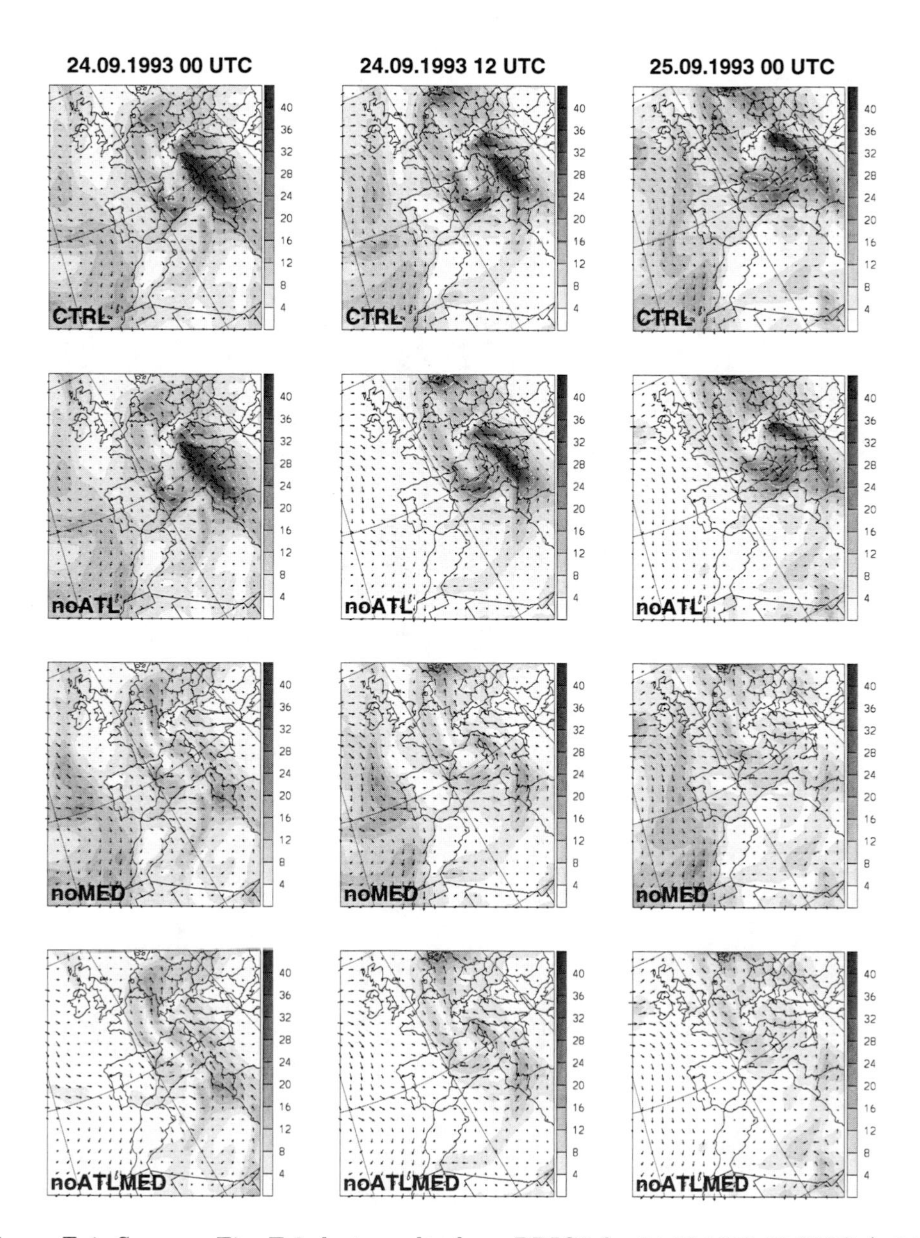

Figure E.4: Same as Fig. E.3, but results from BRIG2 for 24.09.1993 00 UTC (+24h, left-hand panels), 24.09.1993 12 UTC (+36h, middle row panels) and for 25.09.1993 00 UTC (+48h, right-hand panels). See text for more details.

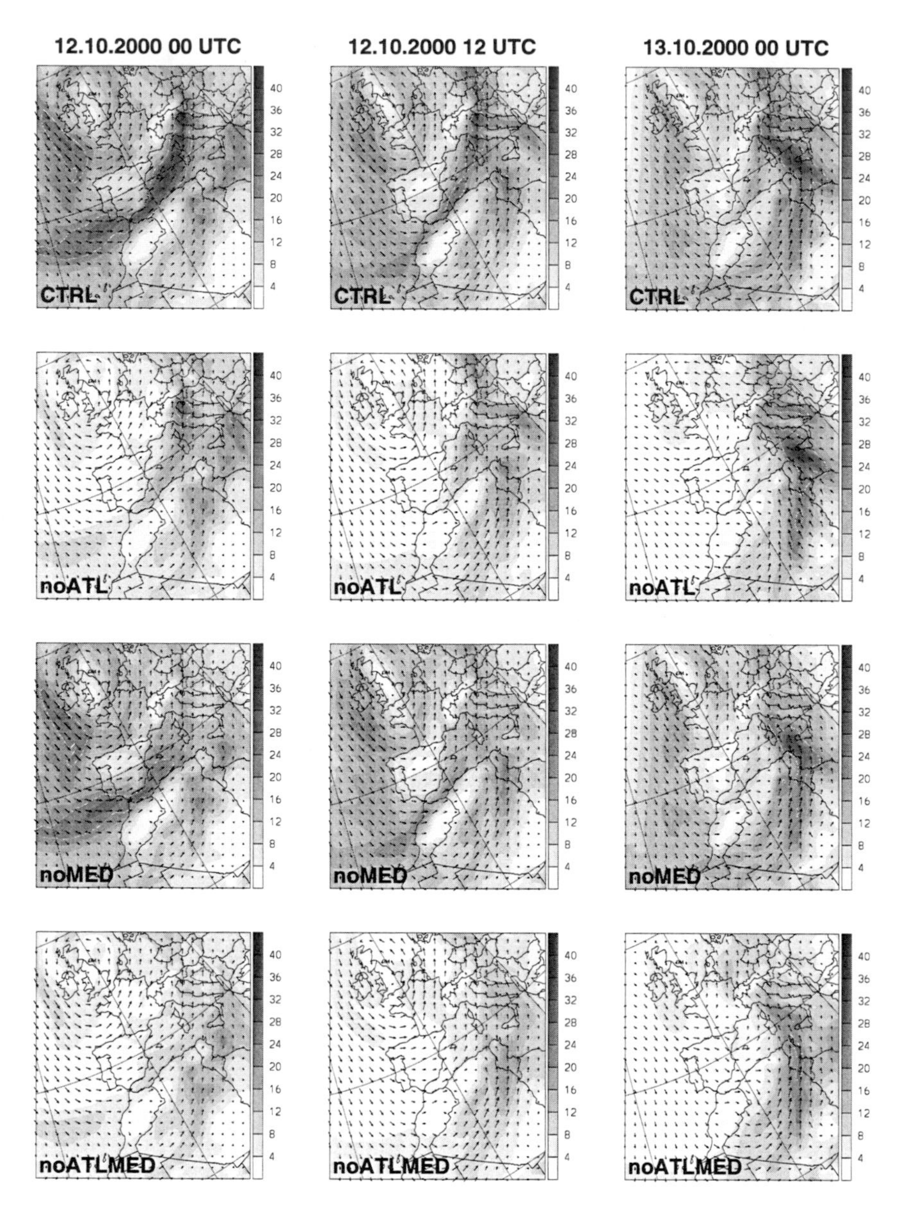

Figure E.5: Horizontal moisture flux vertically integrated between 1000 and 800 hPa (shaded, in kg H_2O/s^3, the 24 kg H_2O/s^3 isoline outlined with a white dashdotted line) and wind vectors at 850 hPa. Results from GONDO1 for 12.10.2000 00 UTC (+24h, left-hand panels), 12.10.2000 12 UTC (+36h, middle row panels) and for 13.10.2000 00 UTC (+48h, right-hand panels). CTRL, noATL, noMED and noATLMED refer to the control simulation and the simulations without evaporation from the Atlantic, the Mediterranean and from both the Atlantic and the Mediterranean. See text for more details.

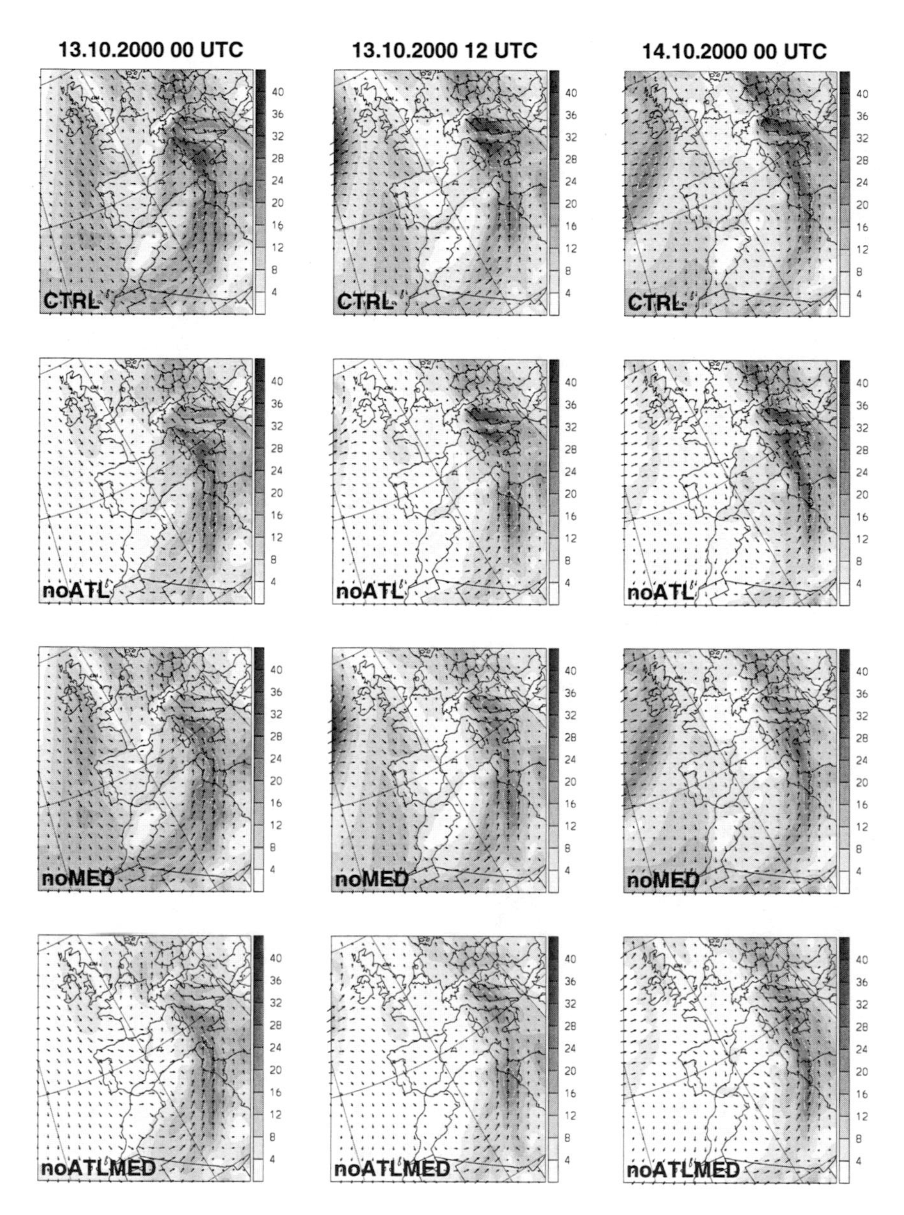

Figure E.6: Same as Fig. E.5, but results from GONDO2 for 13.10.2000 00 UTC (+24h, left-hand panels), 13.10.2000 12 UTC (+36h, middle row panels) and for 14.10.2000 00 UTC (+48h, right-hand panels). See text for more details.

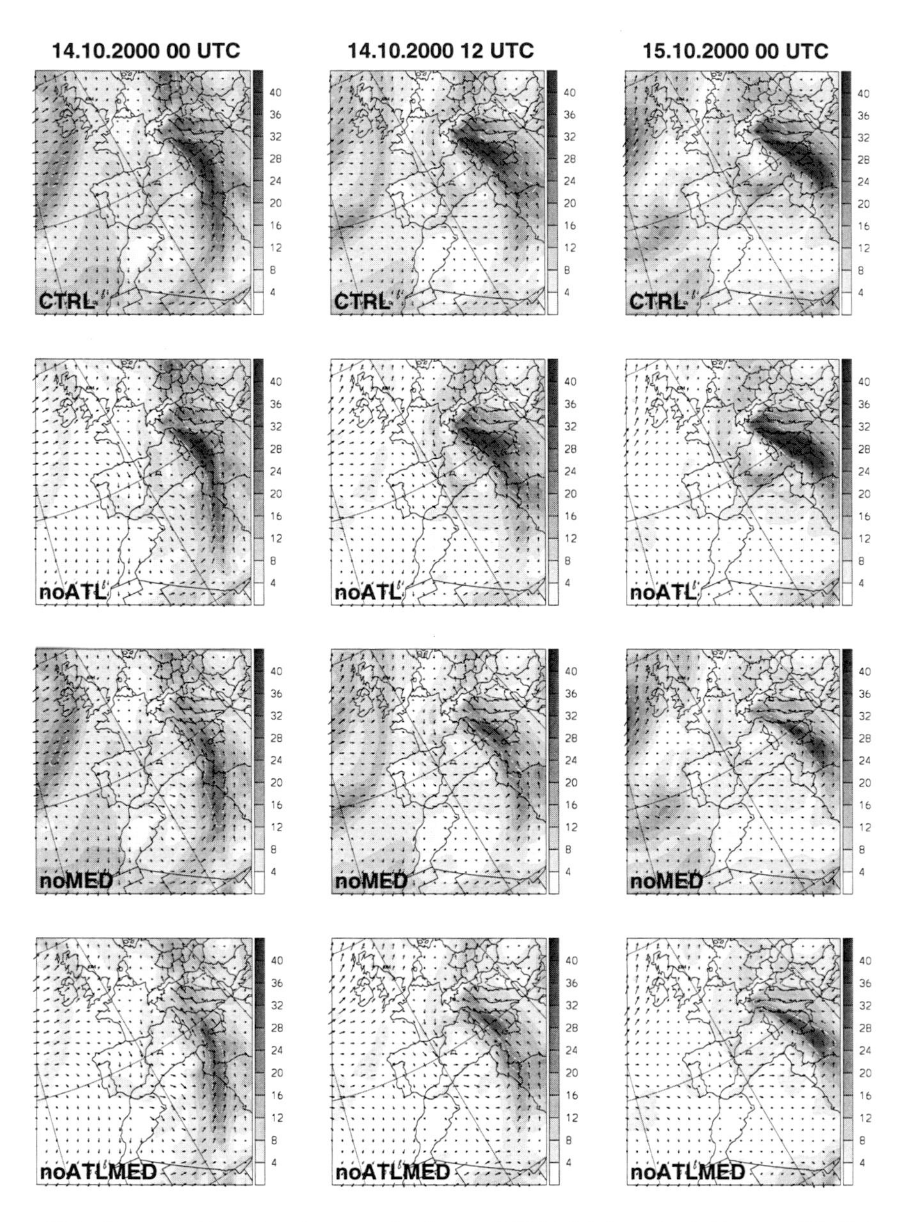

Figure E.7: Same as Fig. E.5, but results from GONDO3 for 14.10.2000 00 UTC (+24h, left-hand panels), 14.10.2000 12 UTC (+36h, middle row panels) and for 15.10.2000 00 UTC (+48h, right-hand panels). See text for more details.

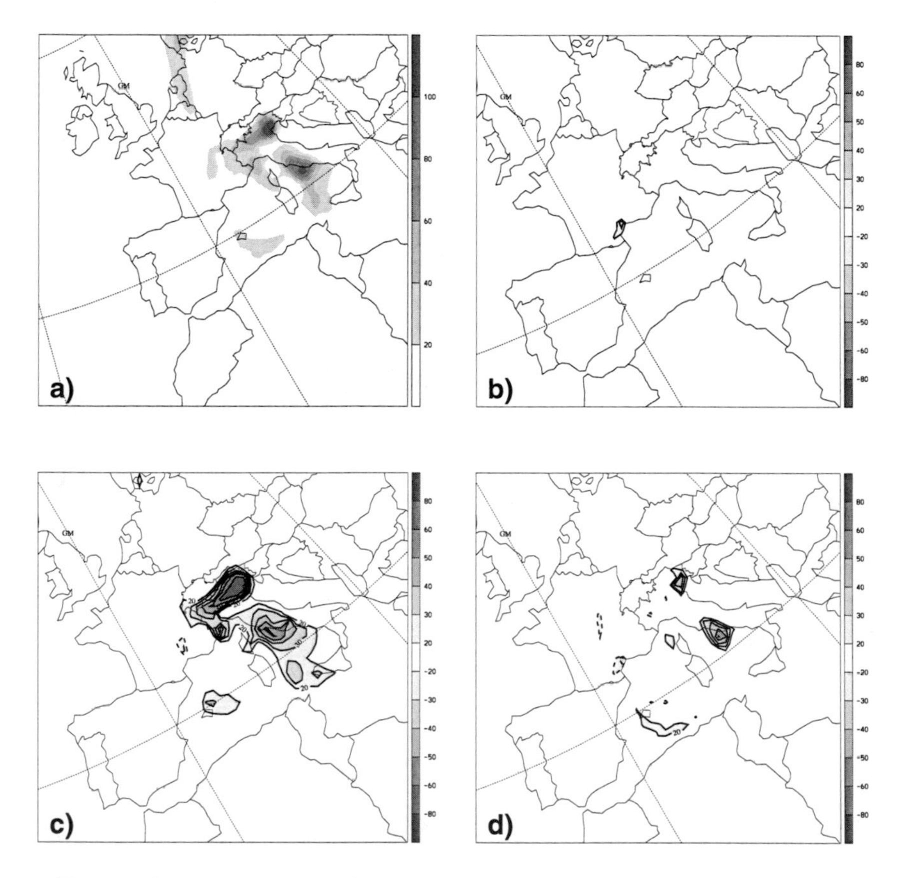

Figure E.8: 24-hour accumulated precipitation in millimeters on 24 September attributed to (b) the Atlantic moisture ($\mathcal{G}_a$), (c) the Mediterranean moisture ($\mathcal{G}_m$), and (d) the interaction between the moisture from the Atlantic and from the Mediterranean $\mathcal{G}_{am}$.(a) shows the 24-hour accumulated precipitation in millimeters on 24 September as given by the CTRL simulation of the BRIG2 experiment.

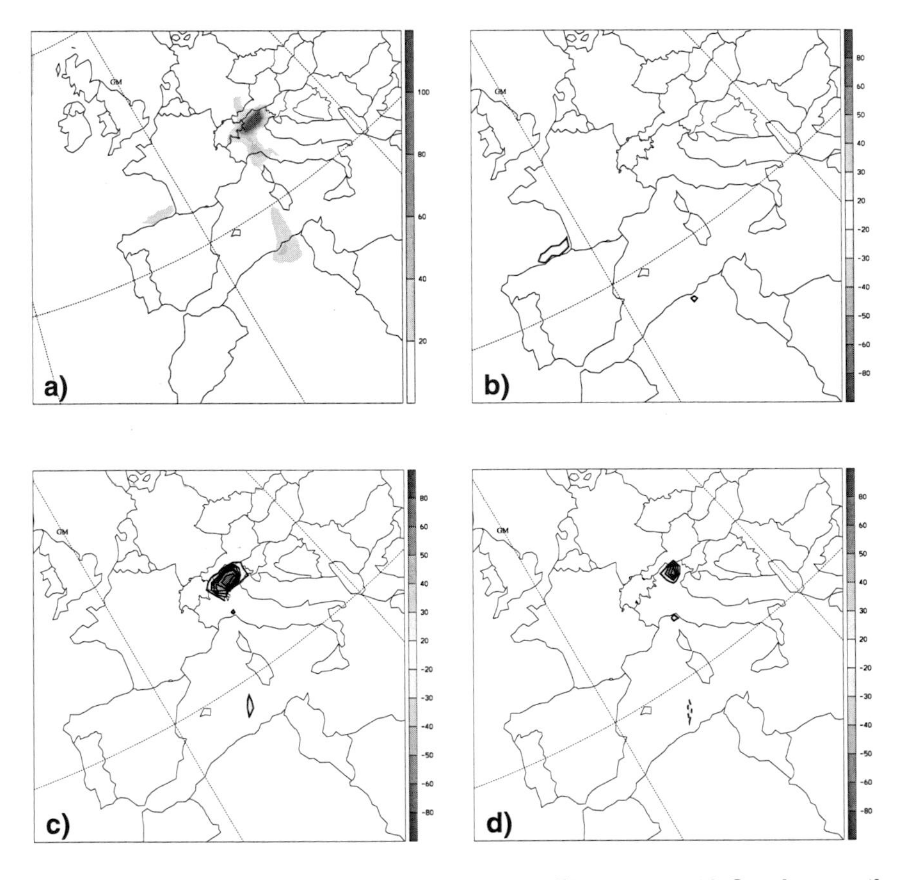

Figure E.9: 24-hour accumulated precipitation in millimeters on 13 October attributed to (b) the Atlantic moisture ($\mathcal{G}_a$), (c) the Mediterranean moisture ($\mathcal{G}_m$), and (d) the interaction between the moisture from the Atlantic and from the Mediterranean $\mathcal{G}_{am}$.(a) shows the 24-hour accumulated precipitation in millimeters on 13 October as given by the CTRL simulation of the GONDO2 experiment.

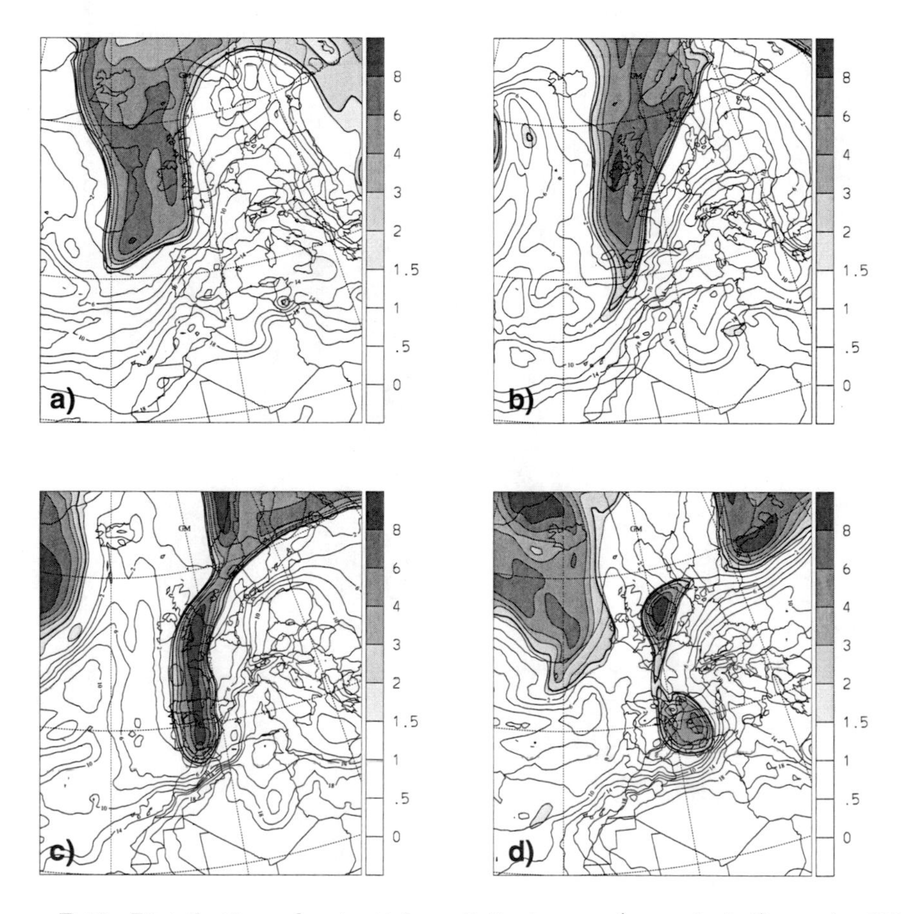

Figure E.10: Distribution of potential vorticity in pvu (grey shaded) on the 320-K isentropic surface and temperature at 800 hPa (solid lines, contours every $2°C$ from $0°C$ to $20°\ C$) at (a) 00 UTC on 21 September 1993, (b) 00 UTC 22 September 1993, (c) 00 UTC 23 September 1993 and (d) 00 UTC 24 September 1993.

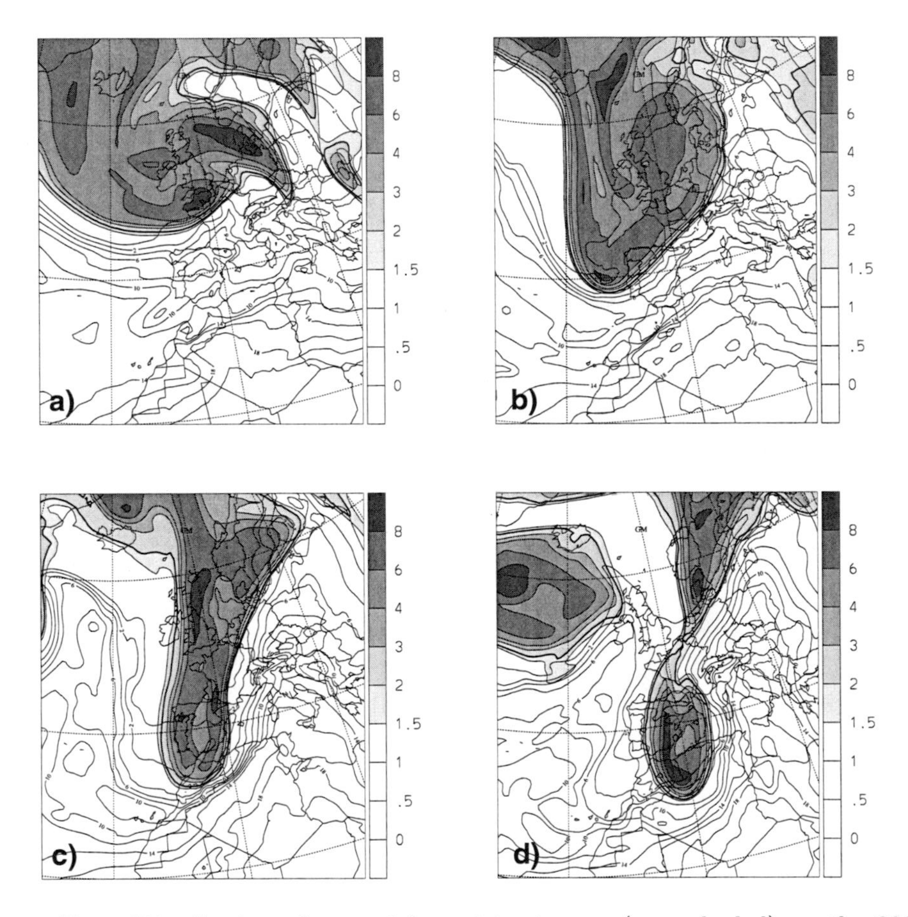

Figure E.11: Distribution of potential vorticity in pvu (grey shaded) on the 320-K isentropic surface and temperature at 800 hPa (solid lines, contours every 2°C from 0°C to 20° C) at (a) 00 UTC on 11 October 2000, (b) 00 UTC 12 October 2000, (c) 00 UTC 13 October 2000 and (d) 00 UTC 14 October 2000.

Bibliography

Alpert, P., M. Tsidulko, S. Krichak, and U. Stein. 1996. A multi-stage evolution of an ALPEX cyclone. *Tellus* **48**, 209–220.

Ambaum, M. H. P., and B. J. Hoskins. 2002. The NAO troposphere-stratosphere connection. *J. Climate* **15**, 1969 – 1978.

Ambaum, M. H. P., B. J. Hoskins, and D. B. Stephenson. 2001. Arctic Oscillation or North Atlantic Oscillation?. *J. Climate* **14**, 3495 – 3507.

Arnal, F. 2003. *Analyse de situations convectives sur les Alpes.* Note de Travail du Groupe de Météorologie à Moyenne Echelle, Météo-France.

Arya, P. S. 1998. *Introduction to micrometeorology.* Vol. 42 International Geophysics Series, R. Dmowska and J. R. Holton eds. Academic Press Inc.

Balasubramanian, G., and M. K. Yau. 1995. Explosive marine cyclogenesis in a three-layer model with a representation of slantwise convection: A sensitivity study. *J. Atmos. Sci.* **52**, 533 – 550.

Baldwin, M. P., and T. J. Dunkerton. 2001. Stratospheric harbingers of anomalous weather regimes. *Science* **294**, 581–584.

Bals-Elsholz, T. M., E. H. Atallah, L. F. Bosart, T. A. Wasula, M. J. Cempa, and A. R. Lupo. 2001. The wintertime Southern Hemisphere split jet: Structure, variablility, and evolution. *J. Climate* **14**, 4191 – 4215.

Barry, R. G., and R. J. Chorley. 2003. *Atmosphere, Weather and Climate - 8th ed.*. Routledge, London.

Benoit, R., and M. Desgagné. 1993. Further non-hydrostatic modelling of the Brig 1993 flash flood event. *MAP Newsletter* **5**, 36–38.

Binder, P., and C. Schär. 1996. *The Mesoscale Alpine Programme: Design Proposal.* Available from MAP Programme Office, c/o Meteoswiss, CH-8044 Zürich.

Bjerknes, J., and H. Solberg. 1922. Life cycle of cyclones and the polar front theory of atmospheric circulation. *Geofys. Publ.* **3 (1)**, 1 – 18.

Blackmon, M. L., J. M. Wallace, N. C. Lau, and S. L. Mullen. 1977. An observational study of the Northern Hemisphere wintertime circulation. *J. Atmos. Sci.* **34**, 1040 - 1053.

Blender, P., K. Fraedrich, and F. Lunkeit. 1997. Identification of cyclone-track regimes in the North Atlantic. *Quart. J. Roy. Meteor. Soc.* **123**, 727–741.

Bluestein, H. 1986. *Fronts and Jet Streaks: A theoretical perspective.* In: Mesoscale Meteorology and Forecasting. Pp. 173–215. American Meteorological Society, Boston. Edited by Peter S. Ray.

Bluestein, H. 1992. *Synoptic-Dynamic Meteorology in Midlatitudes. Volume I: Principles of kinematics and dynamics.* Oxford University Press.

Bluestein, H. 1993. *Synoptic-Dynamic Meteorology in Midlatitudes. Volume II: Observations and Theory of Weather Systems.* Oxford University Press.

Browning, K. 1990. *Organization of clouds and precipitation in extratropical cyclones.* In: Extratropical Cyclones. The Erik Palmén Memorial Volume. Pp. 129–153. American Meteorological Society, Boston. Edited by Chester A. Newton and Eero O. Holopainen.

Buzzi, A., and L. Foschini. 2000. Mesoscale meteorological features associated with heavy precipitation in the southern Alpine region. *Meteorol. Atmos. Phys.* **72**, 131 - 146.

Buzzi, A., N. Tartaglione, and P. Malguzzi. 1998. Numerical simulations of the 1994 Piedmont flood: Role of orography and moist processes. *Mon. Wea. Rev.* **126**, 2369 - 2383.

BWG 2002. *Hochwasser 2000 - Ereignisanalyse/Fallbeispiele, Les crues 2000 - Analyse des événements/Cas exemplaires.* Berichte des BWG, Serie Wasser - Rapports de l'OFEG, Série Eaux Nr.2, available at www.bbl.admin.ch/bundespublikationen.

Chen, B., S. R. Smith, and D. H. Bromwich. 1996. Evolution of the tropospheric split jet over the South Pacific Ocean during the 1986-89 ENSO cycle. *Mon. Wea. Rev.* **124**, 1711 - 1731.

Davies, H. C. 1976. A lateral boundary formulation for multi-level prediction models. *Quart. J. Roy. Meteor. Soc.* **102**, 405 - 418.

Davies, H. C. 1981. An interpretation of sudden warmings in terms of potential vorticity. *J. Atmos. Sci.* **38**, 427 - 445.

Davies, H. C. 1997. *Atmosphärenphysik III: Mesometeorology.* Lecture Notes, ETH Zürch.

Davies, H. C. 1999. *Atmosphärenphysik II - Dynamische Meteorologie.* Lecture Notes, ETH Zürch.

Davies, H. C., and H. Pichler. 1990. Mountain meteorology and ALPEX - an introduction. *Meteorol. Atmos. Phys.* **43**, 3–4.

Davies, H. C., C. Schär, and H. Wernli. 1991. The palette of fronts and cyclones within a baroclinic wave. *J. Atmos. Sci.* **48**, 1666 – 1689.

Davis, C. A. 1992. Piecewise potential vorticity inversion. *J. Atmos. Sci.* **49**, 1397–1411.

Davis, C. A., and K. A. Emanuel. 1991. Potential vorticity diagnostics of cyclogenesis. *Mon. Wea. Rev.* **119**, 1929–1953.

Deser, C. 2000. On the teleconnectivity of the Arctic Oscillation. *Geophys. Res. Let.* **27 (6)**, 779–782.

Doswell, C. A., H. E. Brooks, and R. A. Maddox. 1996. Flash flood forecasting: An ingredients-based methodology. *Wea. Forecasting* **11**, 560–581.

Eady, E. 1949. Long waves and cyclone waves. *Tellus* **1**, 33–52.

Eckert, P., and L. Trullemans. 1999. Breitling Orbiter 3: Meteorological aspects of the balloon flight around the world. *ECMWF Newsletter* **84**, 2 – 6.

Eckhardt, S., A. Stohl, H. Wernli, P. James, C. Forster, and N. Spichtinger. 2004. A 15-year climatology of warm conveyor belts. *J. Climate* **17**, 218–237.

Ertel, H. 1942. Ein neuer hydrodynamischer Wirbelsatz. *Meteor. Z.* **59**, 277–281.

Fehlmann, R., and C. Quadri. 2000. Predictability issues of heavy Alpine south-side precipitation. *Meteorol. Atmos. Phys.* **72**, 223 – 231.

Fehlmann, R., C. Quadri, and H. C. Davies. 2000. An Alpine rainstorm: Sensitivity to the mesoscale upper-level structure. *Wea. Forecasting* **15**, 4 – 28.

Frei, C., and C. Schär. 1998. A precipitation climatology of the Alps from high-resolution rain-gauge observations. *Int. J. Climatol.* **18**, 873 – 900.

Gaberšek, S. 2002. *The dynamics of gap flow over idealized topography.* PhD thesis. University of Washington.

Gheusi, F. 2001. *Analyses eulériennes et lagrangiennes des systèmes convectifs quasi-stationnaires sur les Alpes.* PhD thesis. Thèse de l'Université Toulouse III - Paul Sabatier.

Gheusi, F., and H. C. Davies. 2004. Autumnal precipitation distribution on the southern flank of the Alps: a numerical model study of the mechanism. *Quart. J. Roy. Meteor. Soc.* Accepted.

Gibson, J. K., A. Hernandez, P. Kållberg, A. Nomura, E. Serrano, and S. Uppala. 1997. *ERA description.* ECMWF Re-Analysis Project Report Series 1, ECMWF, Reading, UK.

Gong, D., and S. Wang. 1999. Definition of the Antarctic Oscillation index. *Geophys. Res. Let.* **26 (4)**, 459–462.

Hakim, G. J., L. F. Bosart, and D. Keyser. 1995. The Ohio-Valley wave merger cyclogenesis event of 25-26 January 1978. part I: Multiscale case study. *Mon. Wea. Rev.* **123**, 2663–2692.

Harnik, N., and E. K. M. Chang. 2004. The effects of variations in jet width on the growth of baroclinic waves: Implications for midwinter Pacific storm track variability. *J. Atmos. Sci.* **61**, 23 – 40.

Hartmann, D. L., J. M. Wallace, V. Limpasuvan, D. W. J. Thompson, and J. R. Holton. 2000. Can ozone depletion and global warming interact to produce rapid climate change?. *Proc. Natl. Acad. Sci.* **97**, 1412–1417.

Haynes, P. H., and M. E. McIntyre. 1987. On the evolution of isentropic distributions of potential vorticity in the presence of diabatic heating and frictional or other forces. *J. Atmos. Sci.* **44**, 828–841.

Heck, P. 1999. *European-scale vegetation-climate feedbacks since the time of the Romans. A sensitivity study using a regional climate model.* PhD thesis. Swiss Federal Institute of Technology (ETH). Zürcher Klima-Schriften, Geographisches Institut.

Henderson, J. M., G. M. Lackmann, and J. R. Gyakum. 1999. An analysis of hurricane Opal's forecast track errors using quasigeostrophic potential vortcity inversion. *Mon. Wea. Rev.* **127**, 292–307.

Hortal, M. 1991. *Numerical methods in atmospheric models.* Technical Report, ECMWF.

Hoskins, B. J. 1991. Towards a PV-θ view of the general circulation. *Tellus* **43**, 27–35.

Hoskins, B. J., and K. I. Hodges. 2002. New perspectives on the Northern Hemisphere winter storm tracks. *J. Atmos. Sci.* **59**, 1041 – 1061.

Hoskins, B. J., and N. V. West. 1979. Baroclinic waves and frontogenesis. Part II: Uniform potential vorticity jet flows. *J. Atmos. Sci.* **36**, 1663 – 1680.

Hoskins, B. J., and P. J. Valdes. 1990. On the existence of storm tracks. *J. Atmos. Sci.* **47**, 1854 – 1864.

Hoskins, B. J., I. Draghici, and H. C. Davies. 1978. A new-look at the omega-equation. *Quart. J. Roy. Meteor. Soc.* **104**, 31 – 38.

Hoskins, B. J., M. E. McIntyre, and A. Robertson. 1985. On the use and significance of isentropic potential vorticity maps. *Quart. J. Roy. Meteor. Soc.* **111**, 877–946.

Houze, R. A. 1993. *Cloud Dynamics.* Vol. 53 International Geophysics Series, R. Dmowska and J. R. Holton eds. Academic Press Inc.

Hurrell, J. W. 1995. Decadal trends in the North Atlantic Oscillation and relationships to regional temperature and precipitation. *Science* **269**, 676–679.

IPCC 2001. *Climate Change 2001: The Scientific Basis - Contribution of Working Group I to the IPCC Third Assessment Report.* Available at IPCC Secretariat, c/o WMO, 1211 Geneva 2, Switzerland.

Jones, P. D., T. J., and D. Wheeler. 1997. Extension to the North Atlantic Oscillation using early instrumental pressure observations from Gibraltar and south-west Iceland. *Int. J. Climatol.* **17**, 1433–1450.

Keil, C., H. Volkert, and D. Majewski. 1999. The Oder flood in July 1997: Transport routes of precipitable water diagnosed with an operational forecast model. *Geophys. Res. Let.* **25**, 235 – 238.

Kessler, E. 1969. On the distribution and continuity of water substance in atmospheric circulation models. *Meteor. Monographs* **10**, Americ. Meteor. Soc. Boston, MA.

Kington, J. A. 1999. W. Clement Ley: Nineteenth-Century cloud study and the European jet stream. *Bull. Amer. Meteor. Soc.* **80**, 901–903.

Kleinschmidt, E. 1950. Über Aufbau und Entstehung von Zyklonen (1. Teil). *Met. Runds.* **3**, 1–6.

Knippertz, P. 2003. Tropical-extratropical interactions causing precipitation in northwest Africa: Statistical analysis and seasonal variations. *Mon. Wea. Rev.* **131**, 3069 – 3076.

Knippertz, P., A. H. Fink, A. Reiner, and P. Speth. 2003. Three late summer/early autumn cases of tropical-extratropical interactions causing precipitation in northwest Africa. *Mon. Wea. Rev.* **131**, 116 – 135.

Koch, P. 1999. *A determination of the steering flow of hurricane Andrew (1992) using a piecewise potential vorticity inversion.* Diploma thesis, Swiss Federal Institute of Technology (ETH).

Krau, E. B. 1999. Jet streams revisited. *Bull. Amer. Meteor. Soc.* **80**, 2629.

Krichak, S. O., and P. Alpert. 1998. Role of large scale moist dynamics in November 1-5, 1994, hazardous Mediterranean weather. *J. Geophys. Res.* **103**, 19453 – 19468.

Krishnamurti, T. N. 1961. The subtropical jet stream of winter. *J. Meteorol.* **18**, 172 – 191.

Küttner, J., and co-authors. 1980. *The Alpine Experiment (ALPEX) Design Proposal.* WMO, Geneva.

Lee, S. 1997. Maintenance of multiple jets in a baroclinic flow. *J. Atmos. Sci.* **54**, 1726–1738.

Lee, S., and H.-K. Kim. 2003. The dynamical relationship between subtropical and eddy-driven jets. *J. Atmos. Sci.* **60**, 1490–1502.

Lewis, J. M. 2003. Ooishi's observation viewed in the context of jet stream discovery. *Bull. Amer. Meteor. Soc.* **84**, 357–369.

Lüthi, D., H. C. Davies, A. Cress, C. Frei, and C. Schär. 1996. Interannual variability and regional climate simulations. *Theor. Appl. Climatol.* **53**, 185–209.

Majewski, D. 1991. *The Europa-Modell of the Deutscher Wetterdienst.* Proc. Workshop on Numerical Methods in Atmospheric Models, Vol.2, ECMWF, Reading, United Kingdom.

Massacand, A. C., H. Wernli, and H. C. Davies. 1998. Heavy precipitation on the Alpine southside: An upper-level precursor. *Geophys. Res. Let.* **25**, 1435 – 1438.

Massacand, A. C., H. Wernli, and H. C. Davies. 2001. Influence of upstream diabatic heating upon an Alpine event of heavy precipitation. *Mon. Wea. Rev.* **129**, 2822–2828.

Morgenstern, O., and H. C. Davies. 1999. Disruption of an upper-level PV-streamer by orographic and cloud-diabatic effects. *Contr. Atmos. Phys.* **72**, 173–186.

Murray, R., and S. M. Daniels. 1953. Transverse flow at entrance and exit to jet streams. *Quart. J. Roy. Meteor. Soc.* **79**, 236–241.

Nakamura, H. 1992. Midwinter suppression of baroclinic wave activity in the Pacific. *J. Atmos. Sci.* **49**, 1629 – 1642.

Nakamura, H., and A. Shimpo. 2004. Seasonal variations in the Southern Hemisphere storm tracks and jet streams as revealed in a reanalysis dataset. *J. Climate* **17**, 1828–1844.

Namias, J., and P. F. Clapp. 1949. Confluence theory of the high tropospheric jet stream. *J. Fluid Mech.* **5**, 330 – 336.

Palmén, E., and C. W. Newton. 1969. *Atmospheric Circulation Systems.* Academic Press, New York.

Panetta, R. L. 1993. Zonal jets in wide baroclinically unstable regions: Persistence and scale selection. *J. Atmos. Sci.* **50**, 2073 – 2106.

Pepler, S. J., G. Vaughan, and D. A. Hooper. 1998. Detection of turbulence around jet streams using a VHF radar. *Quart. J. Roy. Meteor. Soc.* **124**, 447 – 462.

Petterssen, S. 1956. *Motion and motion systems.* Vol I, Weather Analysis and Forecasting, McGraw-Hill.

Phillips, N. 1999. Jet streams revisited II. *Bull. Amer. Meteor. Soc.* **80**, 2629–2630.

Pierrehumbert, R. T., and B. Wyman. 1985. Upstream effects of mesoscale mountains. *J. Atmos. Sci.* **42**, 977 – 1003.

Pinto, J. J. G. 2002. *Influence of Large-Scale Atmospheric Circulation and Baroclinic Waves on the Variability of Mediterranean Rainfall.* PhD thesis. Insitut für Geophysik und Meteorologie der Universität zu Köln. Heft 151.

Plumb, R. A., D. W. Waugh, R. J. Atkinson, P. A. Newman, L. R. Lait, M. R. Schoeberl, E. V. Browell, A. J. Simmons, and M. Loewenstein. 1994. Intrusions into the lower stratospheric arctic vortex during the winter 1991-1992. *J. Geophys. Res.* **99**, 1089–1105.

Pomroy, H. R., and A. J. Thorpe. 2000. The evolution and dynamical role of reduced upper-tropospheric potential vorticity in intensive observing period one of FASTEX. *Mon. Wea. Rev.* **128**, 1817 – 1834.

Reale, O., L. Feudale, and B. Turato. 2001. Evaporative moisture sources during a sequence of floods in the Mediterranean region. *Geophys. Res. Let.* **28**, 2085 – 2088.

Reiter, E. R., and A. Nania. 1964. Jet-stream structure and clear-air turbulence (CAT). *J. Applied Meteor.* **3**, 247–260.

Reiter, E. R., and L. F. Whitney. 1969. Interaction between subtropical and polar-front jet stream. *Mon. Wea. Rev.* **97**, 432–438.

Rhines, P. B. 1975. Waves and turbulence on a beta-plane. *J. Fluid Mech.* **69**, 417 – 443.

Riehl, H. 1962. *Jet streams of the atmosphere.* Tech. Rept. No. 32 Dept. Atmospheric Sci., Colorado State Univ., Fort Collins, Colorado.

Riley, D., and L. Spolton. 1981. *World Weather and Climate.* (2nd edn), Cambridge University Press, London.

Rogers, J. C., and H. van Loon. 1982. Spatial variability of sea level pressure and 500 mb height anomalies over the Southern Hemisphere. *Mon. Wea. Rev.* **110**, 1375–1392.

Romero, R., C. Ramis, and S. Alonso. 1997. Numerical simulation of an extreme rainfall event in Catalonia: Role of orography and evaporation from the sea. *Quart. J. Roy. Meteor. Soc.* **123**, 537 – 559.

Rossa, A. M., H. Wernli, and H. C. Davies. 2000. Growth and decay of an extra-tropical cyclone's PV-tower. *Meteorol. Atmos. Phys.* **73**, 139–156.

Rossby, C. G. 1939. Relation between variations in the intensity of zonal circulation of the atmosphere and the displacements of the semi-permanent centers of action. *J. Mar. Res.* **2**, 38–55.

Rotunno, R., and R. Ferretti. 2001. Mechanisms of intense Alpine rainfall. *J. Atmos. Sci.* **58**, 1732 – 1749.

Sanders, F., and B. J. Hoskins. 1990. An easy method for estimation of Q-vectors from weather-maps. *Wea. Forecasting* **5**, 346 – 353.

Schneidereit, M., and C. Schär. 2000. Idealised numerical experiments of Alpine flow regimes and southside precipitation events. *Meteorol. Atmos. Phys.* **72**, 233 – 250.

Schoeberl, M. R., L. R. Lait, P. A. Newmann, and J. E. Rosenfield. 1992. The structure of the polar vortex. *J. Geophys. Res.* **97**, 7859–7882.

Schwierz, C., S. Dirren, and H. C. Davies. 2004. Forced waves on a zonally-aligned jet stream. *J. Atmos. Sci.* **61**, 73–87.

Seibert, P. 1993. Convergence and accuracy of numerical methods for trajectory calculations. *J. Appl. Meteor.* **32**, 558–566.

Seilkopf, H. 1939. *Maritime Meteorologie.* Handbuch der Fliegerwetterkunde. Vol.2, R. Habermehl, Ed. Radetzke.

Shapiro, M. A. 1976. The role of turbulent heat flux in the generation of potential vorticity in the vicinity of upper level jet stream systems. *Mon. Wea. Rev.* **106**, 892 – 906.

Shapiro, M. A. 1980. Turbulent mixing within tropopause folds as a mechanism for the exchange of chemical constituents between the stratosphere and troposphere. *J. Atmos. Sci.* **37**, 994 – 1004.

Shapiro, M. A., and D. Keyser. 1990. *Fronts, jet streams and the tropopause.* In: Extratropical Cyclones. The Erik Palmén Memorial Volume. Pp. 167–191. American Meteorological Society, Boston. Edited by Chester A. Newton and Eero O. Holopainen.

Shapiro, M. A., E. R. Reiter, R. D. Cadle, and W. A. Sedlacek. 1980. Vertical mass constituent and trace constituent transports in the vicinity of jet stream. *J. Atmos. Sci.* **37**, 193 – 206.

Shapiro, M. A., T. Hampel, and A. J. Krueger. 1987. The arctic tropopause fold. *Mon. Wea. Rev.* **115**, 444 – 454.

Shapiro, M. A., W. H., J.-W. Bao, J. Methven, X. Zou, J. Doyle, T. Holt, E. Donall-Grell, and P. Neiman. 1999. *A planetary-scale to mesoscale perspective of the life cycles of extratropical cyclones: The bridge between theory and observations.*. In: The Life Cycles of Extratropical Cyclones. Pp. 139–185. American Meteorological Society. Edited by Melvyn A. Shapiro and Sigbjørn Grønås.

Simmons, A. J. 1991. *Development of a high resolution, semi-Lagrangian version of the ECMWF forecast model.* Numerical methods in atmospheric models, vol. 2, ECMWF.

Sinclair, M. R. 1994. An objective cyclone climatology for the Southern Hemisphere. *Mon. Wea. Rev.* **122**, 2239 – 2256.

Sinclair, M. R. 1997. Objective identification of cyclones and their circulation intensity, and climatology. *Wea. Forecasting* **12**, 595 – 612.

Smith, R. B. 1989. Hydrostatic airflows over mountains. *Adv. Geophys.* **31**, 1–41.

Stein, U., and P. Alpert. 1993. Factor separation in numerical simulations. *J. Atmos. Sci.* **50**, 2107 – 2115.

Stoelinga, M. T. 1996. A potential vorticity-based study of the role of diabatic heating and friction in a numerically simulated baroclinic cyclone. *Mon. Wea. Rev.* **124**, 849 – 874.

Stoelinga, M. T. 2003. Comments on "the evolution and dynamical role of reduced upper-tropospheric potential vorticity in intensive observing period one of FASTEX". *Mon. Wea. Rev.* **131**, 1944 – 1947.

Stohl, A. 2001. A 1-year Lagrangian "climatology" of airstreams in the Northern Hemisphere troposphere and lowermost stratosphere. *J. Geophys. Res.* **106**, 7263 – 7279.

Sutcliffe, R. C. 1939. Cyclonic and anticyclonic development. *Quart. J. Roy. Meteor. Soc.* **65**, 518–524.

Sutcliffe, R. C. 1947. A contribution to the problem of development. *Quart. J. Roy. Meteor. Soc.* **73**, 370–383.

Sutcliffe, R. C., and A. G. Forsdyke. 1950. The theory and use of upper air thickness patterns in forecasting. *Quart. J. Roy. Meteor. Soc.* **76**, 189 – 217.

SwissRe 2003. *Natural Catastrophes and reinsurances.* Available at www.swissre.com.

Thompson, D. W. J., and J. M. Wallace. 1998. The Arctic Oscillation signature in the wintertime geopotential height and temperature fields. *Geophys. Res. Let.* **25**, 1297–1300.

Thorncroft, C. D., B. J. Hoskins, and M. F. McIntyre. 1993. Two paradigms of baroclinic-wave life-cycle behavior. *Quart. J. Roy. Meteor. Soc.* **119**, 17 – 55.

Thorpe, A. J. 2003. Comments on "the evolution and dynamical role of reduced upper-tropospheric potential vorticity in intensive observing period one of FASTEX" - reply. *Mon. Wea. Rev.* **131**, 1948 – 1949.

Thorpe, A. J., and C. H. Bishop. 1995. Potential vorticity and the electrostatics analogy: Ertel-Rossby formulation. *Quart. J. Roy. Meteor. Soc.* **121**, 1477–1495.

Tiedtke, M. 1989. A comprehensive mass flux scheme for cumulus parameterization in large-scale models. *Mon. Wea. Rev.* **117**, 1779–1800.

Trenberth, K. E. 1991. Storm tracks in the Southern Hemisphere. *J. Atmos. Sci.* **48**, 2159 – 2178.

Trüb, J., and H. C. Davies. 1995. Flow over a mesoscale ridge - pathways to regime transition. *Tellus* **47**, 502–524.

Tsidulko, M., and P. Alpert. 2001. Synergism of upper-level potential vorticity and mountains in Genoa lee cyclogenesis - a numerical study. *Meteorol. Atmos. Phys.* **78**, 261 – 285.

Uccellini, L. W. 1990. *Processes contributing to the rapid development of extratropical cyclones.* In: Extratropical Cyclones. The Erik Palmén Memorial Volume. Pp. 81–105. American Meteorological Society, Boston. Edited by Chester A. Newton and Eero O. Holopainen.

Uccellini, L. W., and D. R. Johnson. 1979. Coupling of upper and lower tropospheric jet streaks and implications for the development of severe convective storms. *Mon. Wea. Rev.* **107**, 682 – 703.

Wallace, J. M., and D. S. Gutzler. 1981. Teleconnections in the geopotential height field during the Northern Hemisphere winter. *Mon. Wea. Rev.* **109**, 784–812.

Waugh, D. W., and W. J. Randel. 1999. Climatology of Arctic and Antarctic polar vortices using elliptical diagnostics. *J. Atmos. Sci.* **56**, 1594–1613.

Wernli, H. 1995. *Lagrangian perspective of extratropical cyclogenesis.* PhD thesis. Swiss Federal Institute of Technology (ETH). Dissertation Nr. 11016.

Wernli, H. 1997. A Lagrangian-based analysis of extratropical cyclones. II: A detailed case-study. *Quart. J. Roy. Meteor. Soc.* **125**, 1677 – 1706.

Wernli, H., and C. Schwierz. 2004. A novel method to determine the climatological frequency of cyclones applied to global ERA40 data. *Manuscript in preparation.*

Wernli, H., and H. C. Davies. 1997. A Lagrangian-based analysis of extratropical cyclones. I. The method and some applications. *Quart. J. Roy. Meteor. Soc.* **125**, 467 – 489.

Wernli, H., S. Dirren, M. A. Liniger, and M. Zillig. 2002. Dynamical aspects of the life cycle of the winter storm 'Lothar' (24-26 December 1999). *Quart. J. Roy. Meteor. Soc.* **130**, 405 – 429.

Whitney, L. F. 1977. Relationship of the subtropical jet stream to severe local storms. *Mon. Wea. Rev.* **105**, 398 – 412.

Williams, G. P. 1978. Planetary circulations 1. Barotropic representation of Jovian and terrestrial turbulence. *J. Atmos. Sci.* **35**, 1399 – 1426.

Williams, G. P. 1979a. Planetary circulations 2. The Jovian quasi-geostrophic regime. *J. Atmos. Sci.* **36**, 932 – 968.

Williams, G. P. 1979b. Planetary circulations 3. The terrestrial quasi-geostrophic regime. *J. Atmos. Sci.* **36**, 1409 – 1435.

Williams, G. P. 1988. The dynamical range of global circulations - I. *Climate Dyn.* **2**, 205 – 260.

WMO 2004. Hydrometeorological disaster risk reduction: working together. *WMO Bulletin* **53 (1)**, 15–18.

Yoden, S., M. Shitotani, and I. Hirota. 1987. Multiple planetary flow regimes in the Southern Hemisphere. *J. Meteorol. Soc. Japan.* **65**, 571 – 586.

Zenklusen, E. 2004. *Starkniederschläge auf der Alpensüdseite und ihre Beziehung zu PV-Streamern in der oberen Troposphäre: eine Klimatologie von 1966-99.* Diploma thesis, Swiss Federal Institute of Technology (ETH).

Zhou, S., M. E. Gelman, and A. J. Miller. 2000. An inter-hemispheric comparison of the persistent stratospheric polar vortex. *Geophys. Res. Let.* **27**, 1123–1126.

Zillig, M. 2002. *Dynamics of jet-like flow patterns in the neighbourhood of storms, the tropopause and orography.* PhD thesis. Swiss Federal Institute of Technology (ETH). Dissertation Nr. 14685.

Acknowledgments

The path to the PhD may sometimes be long (three and a half years, although time has passed quite fast...) and paved with new and exciting experiences and discoveries (as well as some frustrations... sometimes). It definitely crosses the paths of other people. Some of them just pass by while some others may become very important. This thesis would never have been possible without the help of some people to whom I would like to express my gratitude here.

My thanks go first to Huw Davies who accepted me as his PhD student. Huw gave me the opportunity to present my work at international conferences and accepted my staying away for two months to recover from my knee surgery (that actually was a great asset for the recovery as these two months were totally icy... you can imagine the situation with crutches...). His position at the head of the Department of Natural Sciences at the ETH unfortunately prevented me from taking fully advantage of his knowledge in atmospheric sciences and incomparable sense of humor. Nevertheless the few (extra-) scientific discussions that were possible have always been great moments.

Thanks to John Gyakum for having accepted to be one of my co-examiner and to have reviewed thoroughly the draft of this thesis, although his position at the head of the Department of the Atmospheric and Oceanic Sciences at McGill University was very time-consuming. Working with John has also been a very motivating and enriching experience.

Thanks to Heini Wernli for having been my second co-examiner and for having been the initiator of the jet study. I will especially remember his enthusiastic view at atmospheric sciences. All the best in Mainz!

My warmest thanks go to the people of the staff of the IACETH for the way they keep the Institute well-working: to Eva Choffat for numerous spontaneous chats, her organizing qualities and some help in the learning of 'Schwiizertüütsch'. To Ruedi Lüthi for his simple way of solving problems, to Peter Isler (I will particularly remember the organization of ICAM) and to Dr. Hans Hirter who introduced me to the (sometimes too-well hidden) 'charms' of Microsoft Windows for the PC-manager job. Thanks a lot to Dani Lüthi for having helped me to solve numerous computer and model problems. My thanks go also to all the (past and present) people of the Dynamic Meteorology Group (and especially to the people of the L6 office) for numerous discussions and help and to all the people of the IACETH (with a special thought to Dottore Stefano, Rafaello and Luo of the L10 office, 'the Place to Be') for all the scientific and extra-scientific nice moments.

I would like to thank the people of the Mesoscale Research Group at McGill University, Prof. Peter Yau, Dr. Ron McTaggart-Cowan and Dr. Jason Milbrandt with whom I shared numerous interesting and enriching discussions during my stay there in October 2003 that substantially helped me in improving the second part of the thesis.

It is a pleasure to thank MeteoSwiss for granting access to the ECMWF data. Mes remerciements vont également aux Drs. Pierre Eckert et André-Charles Letestu de MétéoSuisse Genève pour de très intéressantes discussions ed anché a Igor Giunta di MeteoSvizzera Locarno-Monti per le foto satellite.

I am lucky to have friends with whom I have spent great moments (although the last months of my PhD were characterized by my absence at certain occasions). Merci beaucoup à Corinne (aka Coco) et Florent (Flo), Isabelle (Isa) et Sébastien D. (Didi), Séverine et Steve (Stioufff, bon vent pour ta thèse), Elisabeth et Michel, Sacha (Touille), Antoine F. (El Tonio), Nathalie et Lionel (Toto Kahn), Sylvia and Ron (RMCTC), Marina et Antoine P., Patrick B., François G. (la Franchouille), Philippe et Michelle and to all the friends and people I do not have enough room to write the names of here.

Je tiens particulièrement à remercier ma famille, mes parents et mon frère Jean-Daniel pour toute l'aide et le soutien qu'ils m'ont apportés tout au long de la thèse, ainsi que pour avoir arrangé tous ces petits détails qui facilitent tellement la vie. Et enfin (et surtout), mes plus chaleureux remerciements vont à Karine pour m'avoir soutenu et supporté durant ces trois ans et demi. Cette thèse t'est tout particulièrement dédiée.

Front cover photographs: jet stream over Northern Canada and the Atlantic (left photograph, courtesy of NASA-JSC); aerial view of Gondo, Valais, Switzerland after the landslide of 14 October 2000 (right photograph, courtesy of J.-D. Rouiller, géologue cantonal, e.g. www.crealp.ch).

Curriculum Vitae

Patrick Koch

Ch. de la Crausaz 81
CH-1814 La Tour-de-Peilz
p.koch@alumni.ethz.ch

Born on 2 July 1974
in Vevey (VD)
Nationalities: Swiss and Italian

Education and Professional Training

January 2001 - June 2004 :

PhD thesis at the ETH Zürich, Institute for Atmospheric and Climate Science, in the group for Dynamical Meteorology of Prof. Dr. H. C. Davies.
Title: Novel Perspectives of Jet-Stream Climatologies and Events of Heavy Precipitation on the Alpine Southside.
Dr. Sc. ETH Zürich

May 1999 - December 2000 :

Fellow Scientist at the Office des Finances,
Service des Hospices Cantonaux Vaudois, Lausanne, Switzerland

October 1996 - April 1999 :

2nd cycle university studies in Physics, ETH Zürich
Diploma Thesis in atmospheric physics performed at the Department of Atmospheric and Oceanic Sciences, McGill University, Montreal, Canada.
Title: A determination of the steering flow of Hurricane Andrew (1992) using a piecewise potential vorticity inversion.
Dipl. Phys. ETH

October 1993 - September 1996 :

1st cycle university studies in Physics, EPFL, Lausanne

1990 - 1993 : Gymnase de Burier (VD), Switzerland.
Matura Typus C.
1987 - 1990 : Secondary school in La Tour-de-Peilz (VD), Switzerland.

Publications

- *Novel Perspectives of Jet-Stream Climatologies and Events of Heavy Precipitation on the Alpine Southside* (2004)
P. Koch. PhD Thesis. Swiss Federal Institute of Technology (ETH) Zürich. June 2004.

- Contribution to: *Atmosphere, Weather and Climate, 8th Edition,* (2003), R. G. Barry and R. J. Chorley, Routledge, New York

- *Impacts of the Atlantic and the Mediterranean on a Heavy Precipitation Episode on the Alpine Southside* (2003)
P. Koch and H. C. Davies. Proceedings of the International Conference on Alpine Meteorology (ICAM) and MAP-Meeting, Volume A, pp 73-76. May 2003

- *A determination of the steering flow of Hurricane Andrew using a piecewise potential vorticity inversion.* (1999)
Diploma thesis in atmospheric physics performed at the Department of Atmospheric and Oceanic Sciences, McGill University, Montreal, Canada under the supervision of Professors J. R. Gyakum and M.K. Yau (McGill) and Prof. H. C. Davies (ETH)

- *Treatment of the observations of Stintino and verification with the Swiss Model* (1998)
Practical training at Meteoswiss, Zürich under the supervision of Francis Schubiger

- *February 1997: A study and classification of the perturbation activity over Northern-Atlantic* (1997)
Practical training at the LAPETH, ETH Zürich under the supervision of Prof. H. C. Davies and Alexia Massacand

International Conferences and Workshops

- *An Event-Based Jet-Stream Climatology.*
AGU Joint Assembly 2004. Montreal, Canada, 17-21 May 2004.

- *Impacts of the Atlantic and the Mediterranean on two Heavy Precipitation Episodes on the Alpine Southside.*
  12th Cyclone Workshop. Val Morin, Canada, 21-26 September 2003.

- *Impacts of the Atlantic and the Mediterranean on a Heavy Precipitation Episode on the Alpine Southside.*
  International Conference on Alpine Meteorology and the Mesoscale Alpine Programme (MAP) Meeting 2003. Brig, Switzerland, 19-23 May 2003.

- *Impacts of the Atlantic and the Mediterranean on a Heavy Precipitation Episode on the Alpine Southside.*
  EGS-AGU-EUG Joint Assembly. Nice, France, 6-11 April 2003.

- *Impact de l'Atlantique et de la Méditerranée sur un épisode de fortes précipitations au Sud des Alpes.*
  Atelier de Modélisation de l'Atmosphère (AMA) 2002. Météo France, Toulouse, France, 17-19 December 2002.

- *Summer School on Mountain Meteorology 2002: Modification of Airflow by Mountains.*
  Trento, Italy, 26-30 August 2002.

- *A Global Jet-Stream Climatology.*
  EGS XXVII General Assembly. Nice, France, 22-26 April 2002.

- *Fortes Précipitations au Sud des Alpes et à Gondo (Octobre 2000): Une autre approche.*
  Jahresversammlung der Schweizerischen Gesellschaft für Meteorologie (SGM). Yverdon, Switzerland, 19 October 2001.

Invited Talks

- *The Relevance of the Tropopause Region for Synoptic Weather Systems.*
  Fall Meeting of the Swiss Chapter of the Sigma Xi Scientific Research Society. Bern, Switzerland, 15 November 2003.

- *Impacts of the Atlantic and the Mediterranean on two Heavy Precipitation Episodes on the Alpine Southside.*
  Department of Atmospheric and Oceanic Sciences, McGill University, Montreal, Canada, 1 October 2003.